TOURISM DESTINATION DEVELOPMENT

New Directions in Tourism Analysis

Series Editor: Dimitri Ioannides, E-TOUR, Mid Sweden University, Sweden

Although tourism is becoming increasingly popular as both a taught subject and an area for empirical investigation, the theoretical underpinnings of many approaches have tended to be eclectic and somewhat underdeveloped. However, recent developments indicate that the field of tourism studies is beginning to develop in a more theoretically informed manner, but this has not yet been matched by current publications.

The aim of this series is to fill this gap with high quality monographs or edited collections that seek to develop tourism analysis at both theoretical and substantive levels using approaches which are broadly derived from allied social science disciplines such as Sociology, Social Anthropology, Human and Social Geography, and Cultural Studies. As tourism studies covers a wide range of activities and sub fields, certain areas such as Hospitality Management and Business, which are already well provided for, would be excluded. The series will therefore fill a gap in the current overall pattern of publication.

Suggested themes to be covered by the series, either singly or in combination, include – consumption; cultural change; development; gender; globalisation; political economy; social theory; sustainability.

Also in the series

The Dracula Dilemma
Tourism, Identity and the State in Romania
Duncan Light
ISBN 978-1-4094-4021-5

Emotion in Motion
Tourism, Affect and Transformation
Edited by David Picard and Mike Robinson
ISBN 978-1-4094-2133-7

Tourism Enterprises and the Sustainability Agenda across Europe
Edited by David Leslie
ISBN 978-1-4094-2257-0

Social Media in Travel, Tourism and Hospitality
Theory, Practice and Cases
Edited by Marianna Sigala, Evangelos Christou and Ulrike Gretzel
ISBN 978-1-4094-2091-0

Tourism Destination Development
Turns and Tactics

Edited by

ARVID VIKEN
and

BRYNHILD GRANÅS
UiT - the Arctic University of Norway

LONDON AND NEW YORK

First published 2014 by Ashgate Publishing

2 Park Square, Milton Park, Abingdon, Oxon OX14 4RN
711 Third Avenue, New York, NY 10017, USA

Routledge is an imprint of the Taylor & Francis Group, an informa business

First issued in paperback 2016

British Library Cataloguing in Publication Data
A catalogue record for this book is available from the British Library

The Library of Congress has cataloged the printed edition as follows:
Viken, Arvid.
Tourism destination development : turns and tactics / by Arvid Viken and Brynhild Granas.
    pages cm.
Includes bibliographical references and index.
ISBN 978-1-4724-1658-2 (hardback : alk. paper)—ISBN 978-1-4724-1659-9 (ebook) — ISBN 978-1-4724-1660-5 (epub) 1. Tourism—Marketing. 2. Tourism—Management. 3. Place marketing. I. Title.
  G155.A1V53 2014
  910.68'4—dc23

                                                                          2013043673

ISBN 978-1-4724-1658-2 (hbk)
ISBN 978-1-138-24746-8 (pbk)

Contents

List of Figures and Tables

Figures

Tables

Notes on Contributors

Simone Abram is Reader at the International Centre for Research on Events, Tourism and Hospitality, Leeds Met University. She has published widely in anthropology of tourism and planning, including *Tourists and Tourism* (1997), *Rationalities of Planning* (2002), *Culture and Planning* (2012) and, most recently, *Elusive Promises: Planning in the contemporary world* (2013).

Patrick Brouder holds a PhD in Tourism Studies from Mid Sweden University. He is a post-doctoral researcher at the Department of Tourism Studies and Geography, Mid Sweden University. His research interests include tourism studies, economic geography and rural development. He has published in the Journal of Sustainable Tourism, Tourism Geographies, and the Journal of Heritage Tourism. He was also a guest editor of a special issue of Tourism Planning & Development, which was titled *Tourism Development in Rural Areas*. His PhD thesis is titled *Tourism Development in Peripheral Areas: Processes of Local Innovation and Change in Northern Sweden.*

Jackie Dawson obtained an MA in tourism from the University of Otago, New Zealand and a PhD in geography from the University of Waterloo, Canada. She is currently the Canada Research Chair in Environment Society and Policy in the Department of Geography and Institute for Science, Society and Policy at the University of Ottawa. Her research is on the human- and policy dimensions of climate change, with a particular focus on maritime environments and the Polar Regions.

Anniken Førde is Associate Professor at the Department of Sociology, Political Science and Planning, UiT The Arctic University of Norway. Her research fields are gender studies, place transformation, innovation and tourism. Among her publications are articles and books on cross-disciplinarity, methodology, entrepreneurship and transformation processes.

Brynhild Granås has an MA in sociology from the University of Oslo and a PhD in sociology from UiT The Arctic University of Norway, and is Associate Professor at the Department of Tourism and Northern Studies, UiT The Arctic University of Norway. She has conducted research in place- and urban studies, with a particular focus on the interface between the industrial, cultural, political and material aspects of northern societal development processes. Among her publications are international journal articles in *Geografiska Annaler* and *Acta Borealia*, and her PhD dissertation was titled *The Recalcitrant Manifold. Behind*

Slogans and Headlines of Northern Urban Development. She coedited the book *Mobility and Place. Enacting Northern European Peripheries* (Ashgate 2008) together with J.O. Bærenholdt.

Guðrún Þóra Gunnarsdóttir has an MBA in tourism management from the University of Guelph, Canada, and an MA in comparative literature from the University of Oregon. She is currently a PhD student in tourism studies at the Faculty of Life and Environmental Sciences at the University of Iceland. She is Assistant Professor at the Department of Tourism, Hólar University College, and the head of the department from 1996–2009. She has conducted research in rural tourism development and marketing.

Gunnar Thór Jóhannesson holds an MA degree in anthropology from the University of Iceland and a PhD in Human Geography and Tourism from the University of Roskilde, Denmark. He is currently Assistant Professor at the Department of Life and Environmental Sciences, University of Iceland. He has conducted research in tourism entrepreneurship, innovation, tourism policy and the making of tourism destination. He has also published his work in books and journals, including *Tourist Studies*, *Tourism Geographies* and *Current Issues in Tourism*. He is a co-editor of the edited volume *Actor-Network Theory and Tourism: Ordering, materiality and multiplicity*, published by Routledge in 2012.

Margaret Johnston has a PhD in Geography from the University of Canterbury, New Zealand, and is Professor in the School of Outdoor Recreation, Parks and Tourism at Lakehead University, Canada. Her research examines tourism in the Canadian Arctic, and she worked with a team to undertake community-based research to explore the interplay between tourism change and climate change, focusing on adaptation strategies for communities, the industry and governments. She has published in a variety of northern- and tourism journals, as well as in several collections on northern tourism.

Kari Jæger has an MA in tourism from UiT The Arctic University of Norway and is Associate Professor at the Department of Tourism and Northern Studies, UiT The Arctic University of Norway. Her research interests are in event and tourism studies, with focus on festivals, identity, volunteering, and the connection between tourism and events. Some of her publications are in *Scandinavian Journal of Hospitality and Tourism* and *Event Management*. Her forthcoming PhD thesis (2013) is titled *Understanding Festivals & Events in the European Northern Periphery*.

Eva Kaján has an MA in Cultures and Development Studies from the Katholieke Universiteit Leuven, Belgium, and is current PhD candidate at the Department of Geography at the University of Oulu, Finland. She works as part of the Finnish Research Program on Climate Change (FICCA), funded by the Academy of

Finland. Her research focuses on the relationships between Arctic communities, tourism development and climate change. She has published in international journals such as *Current Issues in Tourism*.

Jenni Lankila has an MSSc in Tourism Research from the University of Lapland, Rovaniemi, Finland, and worked as Research Assistant in METLA (The Finnish Forest Research Institute). She has conducted research in tourism development in rural areas, with a particular focus on the different models of tourism development and their applicability to tourism planning in rural areas. Her master's thesis was titled *Tourism development in national park. Local perspectives on the development of Pallas tourist centre and its surrounding areas and the effects of the development on the operations of local people.*

Dieter K. Müller has a PhD in Social and Economic Geography from Umeå University, and is Professor in Social and Economic Geography at the Department of Geography and Economic History, as well as Dean of the Faculty of Social Science, Umeå University, Sweden. His research addresses second home tourism and the relationship between tourism and regional development in northern areas. Müller has recently co-edited *Tourism in Peripheries* (2007), *Nordic Tourism* (2009), *Polar Tourism – A Tool for Regional Development* (2011) and *New Issues in Polar Tourism* (2013). He is currently the Chairperson of the Commission of Tourism, Leisure and Global Change within the International Geographical Union (IGU).

Torill Nyseth has a PhD in Community Planning from the UiT The Arctic University of Norway, and became Professor in Place Development and Local Planning at the Faculty of Humanities, Social Science and Education, UiT The Arctic University of Norway, in 2009, where she also is the Chair of the research group Place, Power and Mobilities. She has been involved in more than 10 research projects at the European level, particularly in Scandinavia. Her main research topics are place making, network governance and urban planning, and among her publications are international journal articles in *Cities, Planning Theory and Practice, Planning Theory* and *Town Planning Review*.

Jarkko Saarinen is Professor of Human Geography at the University of Oulu, Finland, as well as a Senior Research Fellow at the University of Johannesburg. His research interests include tourism development and sustainability, tourism and climate change and the constructions of nature, local culture and indigeneity in tourism. He is currently the Vice-President of the International Geographical Union (IGU) and Associate Editor in the Journal of Ecotourism. His recent publications include the books, *Tourism and Change in the Polar Regions* (2010, co-edited with C.M. Hall) and *Tourism and Millennium Development Goals* (2013, co-edited with C.M. Rogerson and H. Manwa). He is also a lecturer in tourism at Finnmark University College, Department of Economy and Tourism

Studies. Current interests: cruise-related tourism development; guide performance and tourism experiences.

Ola Sletvold is a lecturer in Tourism at the Department of Tourism and Northern Studies, UiT The Arctic University of Norway. His current research interests are cruise-related tourism development, guide performance and tourism experiences. Previously, he has published on destination development, tourism history, and coastal tourism, and on Viking heritage attractions development. Among his publications is *The Norwegian Coastal Steamer Hurtigruten: Cruising towards Tourism* (in Dowling (ed.) (2006)).

Seija Tuulentie has an MA in Sociology from the University of Tampere, Finland, a PhD in Sociology from the University of Lapland, an Adjunct Professorship of Environmental Sociology at the University of Lapland and is a Senior Researcher of Nature-Based Tourism at the Finnish Forest Research Institute (Metla) in Rovaniemi. Her doctoral dissertation dealt with the rights of the indigenous Sámi people. Over the past 10 years her research interests have been in socially sustainable tourism, second home issues and land use conflicts. She has published international articles, inter alia, in the *Journal of Sustainable Tourism*, the *Scandinavian Journal of Hospitality and Tourism* and in *Forest, Snow and Landscape Research*.

Arvid Viken is Professor in Tourism at UiT The Arctic University of Norway, where he currently is in charge of a PhD study programme in tourism. His research focus is on tourism destinations, northern destinations, tourism industrial development, festivals, place and borders. He has written a series of book chapters and journal articles on tourism, edited several books on tourism in Norwegian, and together with Torill Nyseth, has edited a book titled *Place Reinvention. Northern Perspectives* (Ashgate 2009).

Preface

This book – *Tourism Destination Development. Turns and Tactics* – discusses one of the central phenomena and terms within tourism theory. In this volume tourist destinations are seen in light of the prevailing discourses within the field of tourism studies as well as within society at large, and discussions are based on a series of case studies from the Northern Hemisphere. The book adds to our knowledge about the processes and places that define and prepare for tourism, in addition to tourism as an academic field. It does so by exploring tourism destinations in light of the processes of theming, transformation and politics, as well as presenting conceptual ways of framing the phenomenon. Above all, tourism destinations are understood as dynamic, discursive and spatially connected social fabrics that reflect processes of globalization at one end and local situatedness at the other.

The book is a collection of articles from scholars involved in tourism research who come from different disciplines, but basically from the social science disciplines of sociology, social anthropology, political science and geography. In total, the work originates from the research project, *Chair in Arctic Tourism Research*, at Finnmark University College/Alta (now the University of Tromsø – the Arctic University of Norway/Tromsø and Alta), and involves researchers from Canada, Iceland, Finland, Sweden and Norway. The editors would like to thank all the authors, informants and others who have contributed to the book. The project has been financed by the Norwegian Ministry of Foreign Affairs, to whom the editors and the involved authors are grateful. We also wish to thank Ashgate Publishing for professionally handling all the processes that such a project entails.

Arvid Viken and Brynhild Granås
Alta/Tromsø, June 2013

Chapter 1

Dimensions of Tourism Destinations

Arvid Viken and Brynhild Granås

To let the term 'destination' direct the outline of a book may seem uncontroversial and obviously relevant to the variety of readers of tourism theory. Nevertheless, for social scientists engaged in tourism studies the task has some risks. Firstly, most conceptualizations of the term originate from the scientific field of economics; hence, theoretical elaborations are accompanied – more or less discretely – by economists' basic understanding of ways of approaching the world. Secondly, while many social scientists have become intensively involved in conceptual elaborations and theorizing around the concept (see for example Ringer 1998; Framke 2002; Saarinen 2004; Saraniemi and Kylänen 2011), destination still cannot be considered a stable and nuanced analytical concept for social scientists who approach tourism. Sociologists, social anthropologists, human geographers and political scientists' partaking in such discussions have produced a complex web of alternative, parallel and overlapping suggestions for how to define and approach tourism destinations with regard to research (Britton 1991; Weaver 2000; Bærenholdt 2004; Novelli et al. 2006; Scott et al. 2008; Bramwell 2009), including leaving destination out as an analytical concept altogether. Nevertheless, destination is extensively used within a wider discourse on tourism that also includes practitioners – entrepreneurs, tourists, politicians and planners – who from day to day practice destinations discursively at all societal levels, together with researchers.

Hence, and despite its unstable characteristics, destination is a central and meaningful term in play among all parties concerned with tourism. Within the interdisciplinary collaborative research project behind this book, the conceptual fuzziness in question has proved fruitful in that it has encouraged discussions while striving for a deeper understanding. Our experience has been that its characteristics as a blurring term and topic make destination an analytically potent point of departure for collaborative researchers' 'travel' into current touristic processes, as will be done in this book.

Throughout the book, tourism is understood as a multifaceted and complex phenomenon, for which its economic and business-like marks constitute one aspect among others. Reflected in all chapters is an understanding of tourism destinations as material, sociocultural and dynamic entities. Although variably emphasized by different authors, destinations are understood as continuously constructed through the practices and perceptions of everyday- and touristic life as well as through political and economic decision making processes. Overall, destinations are seen as being embedded in the materialities of landscapes, physical infrastructure and technologies, and in the temporalities of past, present and future. Through the following four chapters (Part I), the authors dedicate attention to conceptual and theoretical discussions and suggest different analytical approaches to destinations, embedded in different social scientific theoretical traditions. The chapters that then follow (Part II-IV) explore emerging and established tourism destinations based on a variety of situated studies from the northern parts of Canada, Iceland, Norway, Sweden, and Finland. The analyses provide opportunities for the reader to take the pulse of current turns and tactics in destination development in the Arctic areas of Europe and Canada.

Thereby, the book also illuminates aspects of today's development of the Arctic as a whole. In social and political terms, the Arctic is often delimited to the areas above the Arctic Circle and includes the northern parts of Russia, Finland, Sweden, Norway, Iceland, Greenland, Canada and the USA (Alaska). Political and research-related discourses on the area often highlight its oceans and ice-covered landscapes and its resources of petroleum, minerals and fish, while connecting it to global questions concerning environmental protection and indigenous rights. However, the Arctic also has extensive and complex patterns of human presence and interaction which involve societal processes such as industrialization, deindustrialization, and urbanization that otherwise mark the nations that stretch into the Arctic.

In addition to this, the northern territories have become an arena for intensified tourism. Increased attention paid to the Arctic from tourism practitioners is in addition to that from political and industrial actors of many kinds, as well as from natural science research milieus and government administrations. Hence, the development of tourism is both determined by and determines the struggle for access and elbow room that otherwise mark the Arctic area. The logic of tourism intersects with the different interests, knowledge systems and rationalities that together make the Arctic the contested area that it is today.

A Situated Study

The studies presented in this book investigate specific destination development processes within the Arctic, and not the Arctic as a destination in itself. One of the questions that can be posed is whether the development that is observed is typical for destinations outside the Arctic. This question touches upon a discussion of what is typical, and if a typical case exists. On the one hand there is a tendency towards seeing what happens in the urban power centres as typical, and in other places as deviant (Gregson et al. 2003; Viken 2012). On the other hand, a very typical travel pattern and motive, is to go to 'other' places. Otherness, newness and exotics are among the tourist buzz-words (Wang 2000). The Arctic has been a place of 'otherness' almost as long as modern tourism has existed. And sometimes a tourism destination in remote areas constitutes a midpoint on the tourism map. This has, for instance, happened to North Cape, the northernmost point of Norway, which was for several periods, most recently in the late 1980s, at the forefront of tourism development in Norway. Here, the first modern, huge and contrived tourist attraction in the country was created. North Cape soon became a model for tourism development elsewhere. It is an attraction and destination on the periphery, but for some years it has been a centre of Norwegian tourism development. However, the question of whether destinations in the Arctic are typical still needs to be addressed.

As the specific destinations explored throughout the book are located in the Arctic area, a remark must be made about a challenge that the authors of this book, along with all researchers who study the north from a northern perspective, have to deal with. The challenge concerns that of exceeding the geographical particulars while communicating research findings from the north to a southern, 'central' and international audience. This challenge is related to a hegemonic touristic discourse about the north, based on narratives of the north produced mainly from the south. Seen from the north, these are narratives that sometimes – based on research approaches that tend to confirm an outsider's preconceptions – underpin homogeneous narratives and myths about the north as for example wild, uncivilized or primitive. With such hegemonic narratives in play, tourism researchers from the north are put under pressure to convince a wider audience about the value of their contributions for concept development and theoretical elaborations within the international field of tourism studies. Looking back at ways of understanding case studies, Lijpart (1977) identifies 'the deviant case study' as one where the case, here being the particular destination, is selected to show an alternative pattern or launch an alternative model. By focusing on Arctic destinations, many will argue that the cases in this book are by definition deviant (Viken 2012). Such an understanding does not correspond with the argument of this book. Our argument is that Arctic destinations also may be typical.

As indicated above, the empirically based chapters of the book explore specific destinations. They do so while accounting for each destination, both in general and specific terms. Even though the theoretical positions taken in regard to philosophy of social science are not stringently coordinated, a shared methodology can be identified in the way all contributions are positioned at a point of intersection between ideographic and nomothetic modes of working. Firstly, all empirical analyses use specific concepts and theories with the aim of contributing to conceptual and theoretical development within the field of tourism studies. Secondly, they provide insight into specific development processes which researchers and students who are engaged in tourism studies elsewhere can learn from in their own empirical and theoretical work. At several points, this common feature in modes of working resembles a situated knowledge approach (Haraway 1996; Hansen and Simonsen 2004; Bærenholdt and Aure 2007). It implies that the ambition to provide knowledge about the particular is not founded on an ideographical way of reasoning. Instead, it is based upon an understanding that emphasizes the partiality of any perspective available to social scientists in their engagement with a complex and changeable world. Similarly, the situated knowledge approach contains no ambitions to establish nomothetic theory, but seeks to establish partial connections which can be identified from the perspective through which the researcher meets the world (Haraway 1996, 264–265).

Most Arctic destinations are not mainstream. None of them are the first destination in most tourists' travel careers. Arctic experiences are based on the area's particular nature, cold climate and experiences of polar nights and midnight sun. The growth seen in Arctic tourism, and particularly in Arctic winter tourism, reflects a global trend of people discovering the aesthetics and values of the cold north. This trend must be understood in relation to the current global interest in the political issue of climate change, the attention this political agenda pays to the Arctic, and the popular symbols that it displays of ice bears, icebergs and the like (Hansson and Norberg 2009). Throughout its history, the core product of Arctic tourism has been that of offering natural scenery for the tourist to gaze at, more or less sheltered on-board ships, coaches, and cars. Today, nature is the major resource in the development of Arctic tourism, but now also as an arena and a resource for sports and activity related tourism, including more focused forms of sightseeing. This trend is part of a general trend towards more specialized forms of leisure (Bryan 1977; Kuentzel and Heberlein 2006) and growing consumer demands for new experiences (Pine and Gilmore 1999; Franklin 2003; Bærenholdt and Sundbo 2007). Thus, the Arctic fits well into a general trend.

Tactics and Turns within the Arctic

The tourist industry is specialized in two ways. First, specialization implies Fordist patterns of development (Ioannides and Debbage 1997) related to that of enhancing productivity through mass tourism. Such quantitatively focused tourism production can be identified in the Arctic, e.g. through an expanding cruise industry within the whole area and through Finnish Christmas tourism in Lapland. Second, and more important within the Arctic, specialization processes also include the development of 'special interest tourism' (Hall and Weiler 1990), and themed travel (Rojek 1993), in destinations normally served by small enterprises, as explored in Part II. The tension between Fordist and alternative patterns of tourism production constitute a backdrop for the formulation of the next section of the book (Part III), where particular focus is placed on current reorientations of well-established destinations. Theming, specialization, standardization and non-standardization of destinations are related to normative questions of what values and visions that tourism development should materialise. The complexity of the phenomenon of tourism implies that tourism development initiatives are interwoven into a manifold of societal processes, such as land use politics, strategic initiatives for the enhancement of employment and economic growth, environmental protection regimes, identity politics and place development processes. Hence, tourism is situated in the midst of processes imbued with politics and power (Cheong and

Miller 2000). The political aspect of destination development is the third area of focus of this book (Part IV).

Themed Destinations

Production and consumption are dynamic and interwoven areas. Changes in consumption follow changes in production, and vice versa. However, the logic of competition encourages product development and therefore the production side often gives the greatest impetus to processes of change. In tourism such product development can comprise everything from transport to attractions and activities on-site, and sometimes also the creation or re-creation of destinations. Among the huge variety of ways in which novel tourism products are pursued is the strategy of theming.

Within the field of tourism studies, the concept of theming is interpreted in a variety of ways which are always concerned with the relatedness of product components. Theming can relate to the narrative tuning of sites of consumption, i.e. the tuning of the variety of parts that comprise a site where consumption takes place. Here, restaurants constitute an archetypical example (Breathworth and Bryan 1999). The degree and consistency of the narrative framing of a site of consumption may vary. For example, and in regard to restaurants, sometimes the menu is the main supporter of a framing narrative while at other times the totality of the product is themed through the use of architecture, interior design, pictures, texts, artefacts and signs within the restaurant premises.

In similar ways, more extensive consumption environments like urban areas and shopping malls can be (re-) shaped as themed environments (Gottdiener 1997). So can tourism attractions and destinations. According to Chang (2000, 36), 'theming' within tourism occurs in at least three ways, i.e. as 'marketing themes' expressed through slogans, catchphrases and marketing material, such as 'theme park and attractions', and as 'place development themes'. According to this definition, a theme chosen for a place will infuse not only strategic thinking and planning, but also product development (Chang 2003), i.e. the development of place as a product, or as a resort development (Flagestad and Hope 2001). One example of themed place development is found in the town of Alta, Norway, where tourists can 'hunt' the Northern Lights, stay at the Northern Lights hotel and visit the Northern Lights Cathedral (from 2013) in a town called 'The City of Northern Lights'. As the observant reader will notice, theming partly overlaps the concept of branding. Within this book, theming is used in categorizing destinations constructed around a core product, such as a sporting activity or a history. Theming indicates that it is not the overall marketing aspects that are in focus throughout the analysis, but how a particular theme can nurture product development and the formation of new destinations.

Branding can be part of a theming process, and theming is closely related to branding. And as with branding there is a risk that the theming will be influenced by trends from other places across the world, with the dynamic of product dedifferentiation in play (Lash 1990; Lash and Urry 1994). Products homogenization across fields and localities is one of the ways globalization is manifested today (Terkenli 2002).

Beach or ski-based destinations and pilgrimages are examples of theme travel, as is the Arctic. But such a huge area also gives room for many types of theming. The Arctic destinations are more or less naturally or culturally themed. Examples of nature-based destination theming are those made around the particular light experiences of the midnight sun or the northern light. In cultural terms, Arctic destinations are often themed in accordance with the culture of indigenous peoples living in the area, like the Inuit or the Sami. Beardsworth and Bryman's (1999) listing of theming classes includes such ethnic marking, in addition to reliquary theming, which is theming related to local rituals or local history. This can also be found within the Arctic, for example in relation to the history of the Arctic expeditions of the nineteenth and early twentieth centuries.

Within the late-modern paradigm of experience tourism, theming strategies are closely related to 'special interest tourism' (Hall and Weiler 1992). People travelling to practice or nourish their hobbies in a new or foreign locality reflect the phenomenon of increased specialization within the leisure field (Kentzel and Heberlein 2006). Becoming more competent, committed, skilled, and specialized in regard to leisure activities hence becomes part of the tourist's agenda. Responses to this trend can be seen at an enterprise level, with companies that relate in fine-tuned and specialized ways to target groups, as well as at a destination level, with specialized activities constituting the core product and main source of the destination theming. A prime example of this kind of specialized activity-based theming of destinations is the off-piste skiing in the Alpine mountain district of Lyngen in Northern Norway. The example also illustrates the extreme sports trend that is unfolding as part of an adventure oriented and risk related Arctic tourism (Mykletun and Gymóthy 2007). Some specialized activities are more strongly conditioned by locally specific natural or cultural features than others. Thus, the destination-activity nexus is manifold.

As with The Sunshine Coast of Spain, with Mecca or with Santiago de Compostela, destinations can be well-established and well-tuned according to a theme long before any actors make strategic initiatives to theme their touristic initiatives or to develop tourism at all. However, the examples of themed destinations presented in Part II of this book exhibit strategic advancements of themed destinations, reflecting the wave of efforts put forth to make tourism a viable industry within the otherwise economically vulnerable communities of the Arctic.

Fordism, Standardization and Antipodes

As sites of production and consumption, destinations are intertwined with the broader trends of capitalist economic development. Such trends are not unambiguous or linear in their character. Similarly, they are not always expressed in clear-cut ways within tourist destinations. The tourism industry of the Arctic has a long history and is rooted in a pre-Fordist and non-industrial economy that can still be identified when approaching certain tourism products and destinations in the area. However, at the same time tourism is based on modern knowledge and technology, including theories of rational organization and conduct. Therefore, a dominant feature of Arctic destinations today is the meeting and mixture between diverging modern and late-modern ideas for economic development, sometimes accompanied by subsequent antagonisms.

One trajectory involved in such meetings and mixtures in Arctic destinations is that of mass-production and standardization associated with terms like 'industrialization' and 'Fordism'. Encouraged by the inner logic of a commercially-based organization as well as the ambitions of planners and politicians, destinations are exposed to the imperatives of growth, i.e. to that of increasing profit through quantitative means in the development of more powerful production units. 'Mass production' is another term used to describe this type of industrial operation and 'mass tourism' is the corresponding term for travel. Destinations have been developed according to this principle for decades, through an economy of scale, hierarchical management, centralised/external management, national/external and regulated finance/ownership, narrow ranges of products, industry-determined products and concentrated production, all features associated with the production modes of Fordism (Ioannides and Debbage 1997, 232; Lafferty and Fossen 2001).

Within tourism, specialized package tours exemplify typical Fordist products, as seen in destinations and resorts all over the world, including the Arctic. Among the Fordist strategies applied to support bigger production units within tourism are horizontal and vertical integration. Vertical integration implies a binding coordination or merging between the vertically identified suppliers within a commodity chain, i.e. for example between airlines, hotels, and tour operators. Sometimes vertical integration entails the model of one company supplying most products; for example, and in regard to ski destinations, some have argued that the most cost-efficient model is when the same company is in charge of lifts, booking, and a significant proportion of the accommodation (Flagestad and Hope 2001). Horizontal integration refers to the merging of companies of the same type, such as when a hotel takes over competing hotels. Examples of both vertical and horizontal integration are found within Arctic destinations, like in Longyearbyen, Svalbard (Chapter 14), where a monopoly situation in the hotel sector has taken shape, or in Lofoten, Norway (Chapter 10) where the presence of an expanding cruise industry leads tourism companies into vertical integration processes.

Parallel with the continuously manifest traces of industrialized and growth-based ideas of economic development, another trajectory present within Arctic destination development is that of late-modern flexible and reflexive production modes, implying a stronger focus on innovative and knowledge-based production (Lash and Urry 1994, 113–123). Destinations in the area are marked by the presence of small and medium sized enterprises (SMEs), often run by well-educated people with lifestyle-based relationships to the products they offer. The lifestyle relationships can be described as twofold: Firstly, the business provides a commercial platform that enables the entrepreneurs to live where – and hence how – they prefer to live and to practice their hobbies and pass on values that they appreciate when performing their business. For some, part of this lifestyle-package can be that of working independently. This latter point leads us to the second lifestyle aspect involved in the SMEs of Arctic destination development, that of a cultural-political protest against the industrialization paradigm of mainstream society, including mainstream tourism development.

Post-Fordist production modes in Arctic tourism are enhanced by the spatial-temporal contextual specificity of the Arctic area, with scarcely populated large landscapes that sometimes constitute unsuitable milieus for large-scale production units. Where large-scale tourism is actually found, another development pattern is that of small-scale development that follows in its wake (Carter 1991). A typical feature would be small local operators that find ways to gain from the tourism traffic generated by big companies. Still another economic strategy behind the post-Fordist flexible production modes are SMEs which concentrate on serving market segments that do not seek mass-products and package tourism, but are willing to pay for the small, the different, the exclusive, and the exotic. Altogether, the reasons for the manifest development of non-industrial and small-scale businesses in Arctic destination development are diverse, as are the meetings and mixtures between such development patterns and other production modes (see Viken and Aarsæther 2013).

Yet another example of overlapping economic strategies is connected to the question of standardization. While the development of the SME-sector described above indicates a product development within tourism marked by differentiation – or diversification – the opposite tendency of dedifferentiation (Lash and Urry 1994, 272) or standardization (Ritzer 1993; 2000) is just as much part of the picture. Ritzer interprets the dynamic of standardization as a legacy of industrial ways of operating, with a focus on rationalization of production through efficiency enhancement, calculability, control and predictability. By making the customers serve themselves – for instance by doing all the booking through an internet-based booking system – efficiency is advanced. By standardizing products, the producer's ability to calculate is supported. Finally, standardization also involves enhancing two-way predictability in the interaction between consumer and producer as well as controlling production, through mechanizing and routinizing it.

The industrializing effects of standardization are evident within today's global tourism, and in relation to small-scale production, with globally applicable standardized expectations of how tourism is supposed to work. Cultural counter reactions to such tendencies are found in anti-modernist trends of seeking enchantment in a seemingly disenchanted world (Ritzer 1999; Franklin 2003) in accordance with a romantic ethic (Campbell 1995). The counter reaction in question manifests itself through celebrations of the past, of myths, of spirituality, and of the ludic aspects of life. As a cultural trend it affects consumer behaviour, particularly within tourism, where 'mistrust towards human spontaneity, drives, impulses and movements that oppose prediction and rational design, has almost completely been replaced by a distrust in the restrained and dominating reason' (Bauman 1993, 33). A re-enchantment trend that materializes this tendency is the one designated by Ritzer as 'Disneyfication' (Ritzer and Liska 1997; Ritzer 1999), exemplifying one among several trends where people's everyday life and environments are 'aestheticized' (Featherstone 1992). According to Ritzer (1999), Disneyfication constitutes a capitalist's answer to the anti-modernist critique described above, designed to secure enduring profit.

As Part III of this book will display, the Arctic is an area where destinations offer possibilities of experiencing the world as re-enchanted, as well as standardized and different, with a mixture of large-scale production units and non-industrial modes of production. This mixture, innovation, growth and regression take place in milieus marked by tensions and conflicts as well as by fruitful concurrences and strong alliances.

Destination Development as Politics

Tourism is political and so are tourist destinations. As an area for development policies, authorities can choose to actively develop tourism as an industry (Reed 1997). Tourism policies are asserted by the national authorities of all Arctic countries today and are also expressed at regional and local levels of politics. At local level, the fact that touristic development initiatives are involved with other policy areas, such as land development, urban regeneration projects and identity politics is immediately apparent. Within regional and national politics, the overlapping and possibly conflicting aspects of tourism development in relation to other policy areas can be kept at an arm's length through rhetoric and systemic specialization. Nevertheless, and as is obvious within the Arctic area, national policy acts related to indigenous rights, environmental protection, industrialization, and tourism sometimes meet up within the localities of tourism destinations and are followed by antagonistic situations.

In addition to this, Reed (1997, 570–572) states that tourism is an arena for allocation policies, i.e. how and where the authorities should spend taxpayers' money or how strongly tourism should be prioritized. The complex characteristics of tourism as a business field, as described above, imply that the expectations of politicians as well as of the population in general are often unstable and sometimes also unrealistic. The outcome is a business that is always included in politicians' speeches, but where money does not always follow the words. Also, tourism is a question of organization, i.e. of how development processes, marketing or assessment projects should be organized, and by whom (Reed 1997). The organization and public involvement in destination management organizations (DMOs) is a central and recurring topic in tourism development in Norway. Even more important is the regulation and governing aspects of organisations and how such aspects relate to the communities that are unavoidably involved in almost any touristic context. Tourism takes place in the midst of existing social environments. It is such pre-established social contexts, which may be in terms of a site, a village, a city, or an urban or a rural area, which are potential objects of transformation that have tourist destinations as their outcomes. Hence, tourism has an influence on people's lives and identities and constitutes a public and political issue.

Antagonisms of destination development may relate to that of governing natural resources, place-making processes, or real estate management, or to questions concerning how the destination should be marketed or organized (Dredge 2010). In such situations, local authorities' methods of coping with tensions between competing stakeholders and networks are particularly relevant. With reference to Forester (1999), Dredge claims that the most likely strategy for local authorities is that of 'critical pragmatism' (Dredge 2006, 563), i.e. being pragmatic enough to get something done but critical enough to worry about the process or result. However, while in one situation pragmatism may mean supporting marginalized actors, it can also imply the opposite.

Over time, the authorities in many places have become well aware of the political implications of destination development, and hence have organized private-public or industry-authority collaboration or partnerships to handle such implications (Hall 1999). Where this is not the case, private development projects still have public implications and are confronted with matters of regulation or governance (Dredge and Pforr 2008). Governance is a way of governing through collaboration or networking, instead of or in addition to that of governing through government. Stoker (1998, 17) describes government as 'activities undertaken primarily or wholly by state bodies' and governance as characterized by 'its focus on governing mechanisms, which do not rest on recourse to the authority and sanctions of governments'. One rationale identified behind governance development lies in the evolving complexity of modern states, where '... no actor has sufficient overview to make the application of particular instruments effective; no single actor has sufficient action potential to dominate unilaterally in a particular governing model.' (Kooiman 1993, 4). Thus, governance refers to regulation, management or steering when a multitude of actors are normally involved, in the process of both decision making and implementation (c.f. Pierre 2000; Pierre and Peters 2000; Kooiman 2003; Kjær 2004). The many governance models described, such as hierarchical governance, self-governance and co-governance (Kooiman 2003), regulation (Clarke 2000), self-regulation (Sanford et al. 2001; Jordan et al. 2003), or co-management (Jentoft 2007), have all been identified within Arctic tourism (Viken 2006).

With reference to Foucault's manifold articulations of power, Cheong and Miller state that 'power is everywhere in tourism' (Cheong and Miller 2000, 372). Hence, power operates within the many and overlapping networks of stakeholders and governance structures of the complex field of tourism. The omnipresent power in question is of a productive kind where power refers to the productivity of all parties involved in the many relations at play (Foucault 1982) within tourism. The network character of power makes the position within networks decisive for the productive power of any actor or group of actors. The tourist gaze exemplifies a relationship imbued with power, as does the guest-host relationship. The asymmetry of resources that actors have within such relationships implies that the relational touristic productivity also can take repressive forms, as seen for example in the organising of the world into rich guests and poorer hosts. The division between a rich and a poor world is part of the structural fundamentals of international tourism (Mowford and Munt 1998). The positioning of centre or periphery is another very common feature of touristic interaction and is obvious in peripheral areas like the Arctic where tourists from urban centres seek to visit remote areas. The asymmetrical relationship in play can also be of a symbolic kind (Granås 2009). In addition to this, destination development becomes a more urgent challenge within peripheries and rural areas than in the urbanized central contexts which are already known and recognized and have developed into tourist destinations by virtue of their central position. As such, destination development strategies reflect structures of hegemony, i.e. of who is in a position to direct the outline of a hegemonic discourse that ascribes places with certain symbolic values. Most often the centre takes on this role. Hence, and in accordance with Foucault's conceptualizations of power, position is more important than knowledge or evidence (cf. Flyvbjerg 1998) when establishing a successful destination. It is the tour operators in the centres that choose destinations or tourism products, even if research-based documentation should indicate that they should go somewhere else.

The discussion of power and politics above opens up a whole series of possible approaches to the analysis of destinations and Part IV of this book elaborates on some of them.

Outline of the Book

Part I, Conceptualizing Destinations, contains three chapters focusing on concepts and theoretical approaches related to its topic. All three dwell on the fact that the way a destination is interpreted and presented depends on the position the analyst holds within the manifold and interdisciplinary field of tourism studies. Arvid Viken (Chapter 2) presents several of the most common approaches to tourism destinations and development, and demonstrates the hegemonic position of the growth paradigm within tourism studies' discourses on destinations. In Chapter 3, Jarkko Saarinen discusses the term in relation to prevailing discourses, particularly emphasising those of regional politics and development, and the implicit changes that occur when destinations are produced and reproduced. Saarinen also discusses the consequences of homogenization and differentiation related to local implementation of globalizing processes. In Chapter 4, Simone Abram writes about how people learn to be tourists through being socialized to and living within discursive frames where tourism exists, but is often taken for granted. She describes processes through which the tourist landscapes are shaped, through practices, politics and planning, and points out the dilemmas in such processes related to power and democracy. In Chapter 5, Brynhild Granås questions the usefulness of basing research on the concept of 'destination'. Her suggestion is, based on a geographical reconceptualization, to replace the term 'destination' with that of 'place', and rather, with destination as a theme of interest, investigate processes within which the place is practiced as a destination. The approach opens up the possibility of identifying the multiple, complex, dynamic and material about power infected processes within which tourism evolves.

In Part II, Catalysing Themed Destinations, three examples are given of how a particular theme has been the basis for a destination development. Guðrún Þóra Gunnarsdóttir and Gunnar Thór Jóhannesson (Chapter 6) show how a history of burning of witches is inscribed into an entrepreneurial process and constitutes a core of a tourism destination development. Gunnarsdóttir and Jóhannesson also show how the people behind this play on networks of ideas, knowledge and local authorities, and how a museum can pave the way for a certain level of small scale tourism. In Chapter 7, Arvid Viken presents the development of a ski resort in Målselv in Northern Norway, with a focus on the types of knowledge involved. The major point of the chapter is to show how scripts and models from tourism literature, but also from the practical field, are applied with an element of the practical knowledge and entrepreneurial skill involved. Furthermore, Kari Jæger and Arvid Viken (Chapter 8) describe the emergence of a sled dog race, the Finnmark Race (*Finnmarksløpet*), as being reciprocal to sled dog tourism development in Finnmark and Norway. Implicit in this is the observation of how this dog race has been a leverage factor in the tourism development of the region in which it takes part, but also that tourism somehow has provided a platform for many sled dog racers. Such relations are characterized as symbiotic.

The four chapters in Part III, Reorienting Destinations, all demonstrate how destinations have to tackle changing environments. Firstly, and based on a relational approach to destination development in Andøy and Målselv in North Norway, Anniken Førde (Chapter 9) focuses on different types of networks as a frame for understanding tourists as well as tourism development. Part of Førde's argument is that tourism becomes a matter of rescue policies for local authorities when other strategies have failed. The chapter also demonstrates how destination development is inscribed in economic discourses related to innovation, growth and branding. Next, Ola Sletvold (Chapter 10) conducts an analysis of the growth of cruise tourism in the Lofoten Islands, a well know destination in North Norway. Sletvold's point of departure is that to be involved with cruise tourism means dealing with standardized mass production. One implication of this is that the local operators have to adapt to a standardized tourism system, but also that this has lots of advantages, such as more business opportunities and higher levels of predictability. In the following chapter (11), Eva Kaján and Jarkko Saarinen problematize the situation in two winter destinations in North Finland, Saariselkä and Kilpisjärvi. The development trajectories of Saariselkä and Kilpisjärvi seem to have been path dependent, and both destinations have faced problems with renewing their products and markets. Kaján and Saarinen also discuss the specific challenges following climate changes that obviously will demand reorientation and renewed

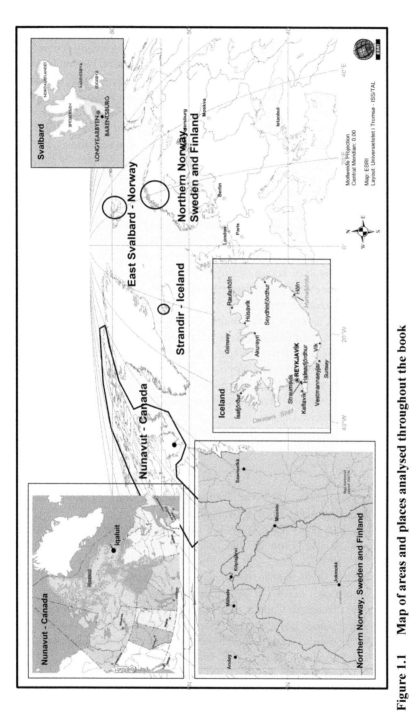

Figure 1.1   Map of areas and places analysed throughout the book

Source: Statens Kartverk 2013

strategies for these destinations. In Chapter 12, Seija Tuulentie and Jenni Lankila analyse how tourism has developed in the municipality of Muonio in North Finland. They show how the development of a small resort, basically a hotel, is inspired by but also locked into the growth strategy of the central authorities, but hindered by environmental constraints because it is located on the border of a national park.

In the final section of the book, Part IV Destination as Politics, the three chapters all articulate how tourism is a matter of politics, but not necessarily an area with offensive public policies. Dieter K. Müller and Patrick Brouder (Chapter 13) point at how tourism is a viable employment sector in North Sweden (Norrbotn), but not so much at the local level of Jokkmokk, a municipality that also includes the national park and the UNESCO world heritage site of *Lapponia*, although this fact adds to the attractiveness of the region. Tourism in the region is growing as a part of the tourism growth discourse that currently also includes northern areas. In Chapter 14, Arvid Viken, Margaret Johnston, Torill Nyseth, and Jackie Dawson discuss management regimes in Nunavut and East Svalbard as responsible governance, focusing on the difference between accountability and responsiveness approaches to tourism governance. They show how accountability, based on research and precaution, has created some problems on Svalbard, whereas a responsive approach on Nunavut has been an obstacle for development.

In the epilogue (Chapter 15), the findings and discussions of all the chapters are brought back into the frame which is presented as the point of departure of the book, i.e. new discussions on concepts and theoretical approaches, theming of destinations, new turns in tourism development, and tourism destinations as products of or arenas for politics. In addition, the final review also unveils some other vital factors of importance in the understanding of tourism destination development, such as knowledge and reflexive stances.

References

Baldachino, G. (ed.) (2006), Extreme Tourism. Lessons from the World's Cold Water Islands (New York: Elsevier Science).

Beardsworth, A. and Bryman, A. (1999), 'Late Modernity and the Dynamics of Quasification: The Case of the Themed Restaurant', The Sociological Review 47:2, 228–257.

Bramwell, B. (2011), 'Governance, the State and Sustainable Tourism: A Political Economy Approach', *Journal of Sustainable Tourism* 19:4/5, 459–477.

Britton, S. (1991), 'Tourism, Capital and Place: Towards a Critical Geography of Tourism', *Environment and Planning D: Society and Space* 9:4, 451–478.

Bryan, H. (1977), 'Leisure Value Systems and Recreational Specialization: The Case of Trout Fishermen', *Journal of Leisure Research* 9:3, 174–187.

Bærenholdt, J.O. and Sundbo, J. (eds) (2002), *Opplevelsesøkonomi. Produktion, forbruk, kultur* (Frederiksberg: Forlaget Samfundslitteratur).

Bærenholdt, J.O. and Aure, M. (2007), 'Tværfaglighed og sammenligning på tværs mod nord', in Nyseth et al. (eds).

Carter, F.W. (1991), 'Czechoslovakia', in Hall (ed.).

Chang, T.C. (2000), 'Theming Cities, Taming Places: Insights from Singapore', *Geografiske Annaler* 82 B:1, 35–53.

Cheong, S.M. and Miller, M.L. (2000), 'Power and Tourism: A Foucauldian Observation', *Annals of Tourism Research* 27:2, 371–390.

Clarke, M. (2000), *Regulation. The Social Control of Business between Law and Politics* (Basingstoke: Macmillan Press).

Dredge, D. (2006), 'Networks, Conflict and Collaborative Communities', *Journal of Sustainable Tourism* 14:6, 562–581.

—— (2010), 'Place Change and Tourism Development Conflict: Evaluating Public Interest' *Tourism Management* 31:1, 104–112.

Dredge, D. and Pforr, C. (2008), 'Policy Networks and Tourism Governance', in Scott et al. (eds).

Featherstone, M. (1992), *Consumer Culture and Postmodernism* (London: Sage).

Flagestad, A. and Hope, C.A. (2001), 'Strategic Success in Winter Sports Destinations: A Sustainable Value Creation Perspective', *Tourism Management* 22:5, 445–461.

Flyvbjerg, B. (1998), *Rationality and Power: Democracy in Practice* (Chicago: University of Chicago Press).

—— (2006), 'Five Misunderstandings about Case-Study Research', *Qualitative Inquiry* 12:2, 219–245.

Forester, J.F. (1999), *The Deliberative Practitioner: Encouraging Participatory Planning Processes* (London: MIT Press).

Foucault, M. (1982), 'The Subject and Power', *Critical Inquiry* 8:4, 777–795.

Framke, W. (2002), 'The Destination as a Concept: A Discussion of the Business-Related Perspective versus the Socio-Cultural Approach in Tourism Theory', *Scandinavian Journal of Hospitality and Tourism* 2:2, 92–107.

Franklin, A. (2003), *Tourism: An Introduction* (London: Sage).

Fryer, P. et al. (eds) (2010), *Encountering the Changing Barents: Research Challenges and Opportunities* (Juansuu: Arctic Centre Reports – Arktisen keskuksen tiedotteita).

Gottdiener, M. (1995), *Postmodern Semiotics. Material Culture and the Forms of Postmodern Life* (Oxford: Blackwell).

Gregson, N.A., Simonsen, K. and Vaiou, D. (2003), 'Writing (Across) Europe: On Writing Spaces and Writing Practices', *European Urban and Regional Studies* 10:1, 5–22.

Granås, B. (2009), 'Constructing the Unique – Communicating the Extreme. Dynamics of Place Marketing', in Nyseth and Viken (eds).

Hall, C.M. (1999), 'Rethinking Collaboration and Partnership: A Public Policy Perspective', *Journal of Sustainable Tourism* 7:3/4, 274–279.

Hall, C.M. and Weiler, B. (1992), *Special Interest Tourism* (Bellhaven Press).

Hall, D.R (ed.) (1991) *Tourism and Economic Development in Eastern Europe and the Soviet Union* (London: Belhaven).

Hansen, F. and Simonsen, K. (2004), *Geografiens videnskapsteori. En introducerende diskussion* (Frederiksberg: Roskilde Universitetsforlag).

Hansson, H. and Norberg, C. (2009), 'Revisioning the Value of Cold', in Hansson and Norberg (eds).

Hansson, H. and Norberg, C. (eds) (2009) *Cold Matters. Cultural Perceptions of Snow, Ice and Cold* (Umeå: Umeå University and the Royal Skyttean Society).

Haraway, D. (1996), 'Situerte kunnskaper. Vitenskapsspørsmålet i feminismen og det partielle perspektivets forrang', in Sæter (ed.).

Ioannides, D. and Debbage, K. (1997), 'Post-Fordism and Flexibility: The Travel Industry Polyglot', *Tourism Management* 18:4, 229–241.

Jordan, A., Rüdiger Wurzel, K.W. and Zito, A.R. (2003), '"New" Instruments of Environmental Governance. Patterns and Pathways of Change', in Jordan et al. (eds).

Jordan, A. et al. (eds) (2003), *'New' Instruments of Environmental Governance. National Experiences and Prospects* (London: Frank Cass).

Kjær, A.M. (2004), *Governance* (Cambridge: Polity Press).

Kooiman, J. (1993), *Modern Governance: New Government-Society Interactions* (London: Sage).

Kooiman, J. (2003), *Governing as Governance* (London: Sage Publications).

Kuentzel, W.F. and Heberlein, T.A. (2006), 'From Novice to Expert? A Panel Study of Specialization Progression and Change', *Journal of Insure Research* 38:4, 496–512.

Lafferty, G. and Fossen, A. (2001), 'Integrating the Tourism Industry: Problems and Strategies', *Tourism Management* 22:1, 11–19.

Lash, S. (1990), *The Sociology of Postmodernism* (Routledge, London).

Lash, S. and Urry, J. (1994), *Economies of Signs and Space* (London: Sage Publications).

Lijphart, A. (1975), 'The Comparable-Cases Strategy in Comparative Research', *Comparative Political Studies* 8:2, 158–77.

Mowforth, M. and Munt, I. (1998), *Tourism and Sustainability* (London: Routledge).

Mykletun, R. and Gymóthy, S. (2007), 'Adventure Tourism: Den grænseoverskridende leg med risiko', in Bærenholdt and Sundbo (eds).

Novelli, M., Schmitz, B., and Spencer T. (2006), 'Networks, Clusters and Innovation in Tourism: A UK Experience', *Tourism Management* 27:6, 1141–1152.

Nyseth, T. et al. (eds) (2007) *I disiplinenes grenseland. Tverrfaglighet i teori og praksis* (Bergen: Fagbokforlaget).

Nyseth, T. and Viken, A. (eds) (2009) *Place Reinvention: Northern Perspective* (London: Ashgate).

Pierre, J. (2000), *Debating Governance* (Oxford: Oxford University Press).

Pierre, J. and Guy Peters, B. (2000), *Governance, Politics and the State* (London: Macmillan Press).

Pine, B.J. and Gilmore, J.H. (1999), *The Experience Economy: Work is Theatre and Every Business a Stage* (Boston, MA: Harvard Business Press).

Reed, M.G. (1997), 'Power Relations and Community-Based Tourism Planning', Annals of Tourism Research 24:3, 566–591.

Rojek, C. (1993), 'Disney Culture', Leisure Studies 12:1, 121–135.

―――― (1993), *Ways of Escape: Modern Transformations in Leisure and Travel* (London: Macmillan).

Rojek, C. and Urry, J. (eds) (1998), *Touring Cultures: Transformations in Travel and Theory* (London: Routledge).

Ringer, G. (1998), 'Introduction', in Ringer (ed.).

Ringer. G. (ed.) (1998), *Destinations: Cultural Landscapes of Tourism* (London: Routledge).

Ritzer, G. (1998), *The Mcdonaldization Thesis* (London: Sage Publications).

―――― (1999), *Enchanting a Disenchanted World. Revolutionizing the Means of Consumption* (Thousand Oaks: Pine Forge Press).

Ritzer, G. and Liska, A. (1998), '"McDisneyization" and "Post-Tourism": Complementary Perspectives on Contemporary Tourism', in Rojek and Urry (eds).

Saarinen, J. (2004), 'Destinations in Change. The Transformation Process of Tourist Destinations', *Tourist Studies* 4:2, 161–179.

Sanford, E.G. and Kimber, C. (2001), 'Redirecting Self-Regulation', *Journal of Environmental Law* 13:2, 158–185.

Saraniemi, S. and Kylänen, M. (2011), 'Problematizing the Concept of Tourism Destination: An Analysis of Different Theoretical Approaches', *Journal of Travel Research* 50:2, 133–143.

Scott, N., Cooper, C. and Baggio, R. (2008), 'Destination Networks: Four Australian Cases', *Annals of Tourism Research* 35:1, 169–188.

Scott, N. et al. (eds) (2008), *Network Analysis and Tourism: From Theory to Practice* (Clevedon: Channel View Publications).

Stoker, G. (1998), 'Governance as Theory', *International Social Science Journal* 50:155, 17–28.

Sæter, G. (ed.) (1996), *HUN – en antologi om kunnskap fra kvinners liv* (Oslo: Spillerom).

Terkenli, F.S. (2002), 'Landscapes of Tourism: Toward a Global Cultural Economy of Space?', *Tourism Geographies* 4:3, 227–254.

Viken, A. (2006), 'Svalbard', in Baldachino (ed.).

―――― (2010), 'Academic Writing about Arctic Tourism: Othering of the North', in Fryer et al. (eds).

Viken, A. and Aarsæther, N. (2013), 'Transforming an Iconic Attraction into a Diversified Destination: The Case of North Cape Tourism', *Scandinavian Journal of Hospitality and Tourism* 13:1, 38–54.

Wang, N. (2000), *Tourism and Modernity: A Sociological Analysis* (Oxford: Pergamon Press).

Weaver, D.B. (2000), 'A Broad Context Model of Destination Development Scenarios', *Tourism Management* 21:3, 217–224.

PART I
Conceptualizing Destinations

Chapter 2
Destinations Discourses and the Growth Paradigm

Arvid Viken

Introduction

In tourism studies, the destination is as central as the tourist. To travel and stay away from home is the essence of tourism; as tourists people may go to one or several destinations. The term 'destination' is a 'predetermined end', or in the terms of tourism literature; 'a place or point aimed at' (Haldrup 2001). Bornhost and colleagues (2010, 572) define a destination as 'a geographical region, political jurisdiction, or major attraction, which seeks to provide visitors with a range of satisfying to memorable visitations experiences'. However, still it is a rather fuzzy term, with regard to both spatial aspects and similar concepts, and with regard to the role of the tourist in the production of a destination (Haldrup 2001). There are a number of terms which are used more or less synonymously with 'tourist destination', such as 'tourist places' (Bærenholdt et al. 2004), 'tourism areas' (Butler 1980), 'tourist regions' (Saarinen 2004), and 'tourist attraction system' (Leiper 1990a). The aim of this chapter is to show that the term – as it is applied in the tourism literature – is not neutral, but is influenced by disciplines, discourses and ideologies, and particularly by what is called here the 'growth paradigm', where economic growth is the key to economic development, as described by Schumpeter (1934). The focus on economic growth, called 'growth fetishism' by Hamilton (2003), is the core of the neoliberal and the market economy thinking. Hamilton claims that this has transferred 'political authority from the state to the private market' (ibid., 17), referring to the positions and power of international corporations, and modern governance regimes. Neoliberalization, along with the idea of globalization, has had a strong effect on analyses of contemporary tourism (Beaumont and Dredge 2010). Higgins-Desbiolles (2006, 1194), referring to globalization, argues that 'the tourism sector is very important in these processes because the consumption of the tourism experience is a key 'growth' sector in many contemporary economies'. As will be demonstrated in this chapter this is also a mantra within academic accounts of destinations and tourist places. The mantra will here generally be referred to as the 'market economy' philosophy, or the 'growth paradigm'; a paradigm being loosely understood as a way of thinking.

Although destinations have a materiality related to places or regions, they are primarily socially constructed. When 'destination' is used about a place, it implies a way of looking at a place which is very different to looking upon it as a site for grazing or dwelling. The particular ways of thinking about places mentioned here can be seen as a collection of discursive practices. The discourse of tourism transforms travellers into tourists, and places into destinations. It is a powerful discourse – the media today is filled with travelling suggestions and advertisements for places to go. Affiliated to this is the academic discourse about destinations. In conceptualizing 'destination', terms like 'development', 'growth', 'branding', and 'marketing' tend to be used. According to Saarinen (2004), there are strong discourses within the tourism realm that force thinking about destinations in the direction of standardization and growth. Although they are strongly influenced by the economical perspective, 'destination' analyses are constructed within the disciplines to which the scholars belong (Bornhorst et al. 2010). According to Higgins-Desbiolles (2006, 1198) there is a major split between two perspectives; tourism as an industry, and as a social force. The industry approach considers issues like profit, market, industrialization, standardization and growth. As a social force the focus is on issues like governing, community, social needs, social concerns, and welfare. Framke (2002), in a discussion of theoretical approaches to destinations, identifies a major split between economic and social science based approaches, although he also finds that in the last area the term 'destination' is applied less often. In this chapter it will be shown that 'destination' is a term influenced by the growth perspective, and its underlying ideologies, even in analyses which have non-commercial points of departure. An overview of major approaches to destination development will be given, based on a review of accounts in the field.

The chapter starts with a presentation of discourse analysis. Seven approaches to destination and destination development are presented, including some which are strongly rooted in the growth field, and some which represented an alternative, or even an oppositional approach: destinations as industrial development; destinations as networks and stakeholder groups; destinations as a strategic matter; destinations as branding and management of symbols; destinations as places and communities; destinations as attraction systems and ideologically based tourism destination models. In this discussion, examples will also be given of how the discourses are inscribed in development processes, mostly with examples from Norway. In a subsequent section, destination development is discussed in relation to globalization, contemporary trends and fads, and as a reflexive matter.

Discourse Analyses of Tourism Discourses

Discourse analysis deals with power relations. Its aim is to detect which actors, narratives, and ideologies dominate an academic or public field. One of those who have discussed the relations between discourses, power, and dominance is van Dijk (1993), who is inspired by scholars as Weber, Foucault and Gramsci. According to van Dijk, dominance is seldom total, but most often restricted to a particular area, and it is contested. He states that:

> If the minds of the dominated can be influenced in a such way that they accept dominance, and act in the interest of the powerful out of the own free will, we use the term *hegemony* (Gramsci 1971; Hall et al. 1977). One major function of dominant discourse is precisely to manufacture such consensus, acceptance and legitimacy of dominance (Herman and Chomsky 1988). (van Dijk 1993, 255)

Van Dijk (ibid., 255) also argues that 'power and dominance are usually *organised* and *institutionalized*', and that dominance implies a *hierarchy of power.* One institution in action is a scientific discipline. Certain people control the most important positions and arenas for the academic discussions. These include university positions, the knowledge required for these, the research programs and funds, the journals, the rating of journals, the language of the journals, and the publishing channels as such. The most important is probably that a discipline or a group within a discipline can be seen as a 'power elite' with 'special access to discourse: they are literally the ones who have most to *say*' (ibid., 255). Van Dijk (1993, 257) also discusses the way power exists within a discourse: basically through 'social cognition' and 'mind management'. Writing that supports a particular discourse or ideology is managing our minds in a particular direction.

A way of identifying discourses and their ideological implications is to make a discourse analysis of theoretical outlines – in this case tourism destination theories. In doing so, this analysis will lean on the approaches of Laclau and Mouffe (1985), and Fairclough (1992). A discourse, according to Laclau and Mouffe (1985), consists of the following elements:

> we will call *articulation* any practices establishing a relation among elements such that their identity is modified as a result of the articulatory practice. The structured totality resulting from the articulatory practice, we will call *discourse*. The differential positions, insofar as they appear articulated within a discourse, we will call moments. By contrast, we will call *element* any differences that is not discursively articulated. (Laclau and Mouffe, 1985, 105; emphasis in the original)

Moments are the vital elements of a discourse; signs that when they are combined, fixate their meaning on a particular discourse. A discourse is more or less locked, but is often defined in relation to other discourses. Thus, elements from other discourses may be included. Therefore, most discourses tend to be hybrids, including elements from the discourses with which they compete. According to Laclau and Mouffe (1985, 112), discourses try to 'dominate a field of discursivity ..., to construct a centre', which they call a 'nodal point'. According to Philips and Jørgensen (2000), a discourse is structured by, and given meaning through, one or several nodal points. Nodal points are articulations of so called floating signifiers, elements that in principle can support different discourses. Although fixation is an inherent motive in most discourses, they are also dialectical and dynamic, and can always be opened up and given new directions.

Discourse analysis has been applied to scholarly based disciplines and discourses before; for example in many feminist studies (see i.e. Aitchinson 2000). Disciplines are filled with signifiers supporting a particular stance, and reflect particular world views. The aim of this analysis will be to detect nodal points and underlying discourses and ideologies. The most successful discourses are those which, being implicit, are taken for granted. In discourse analysis the process leading towards this situation is labelled 'naturalization'. 'Through the process of naturalization individual opinions become 'common sense' and appear to lose their 'partial' character, reappearing as 'neutral' or 'true'" (Vázquez 2006, 59). Discourses are both subject to, and a way taken-for-grantedness enacts. According to Fairclough (1989), naturalization is one of the most effective ways to legitimize power, and Vazquez (op.cit.) shows how this can be done through communicating the discourses through daily life practices and language. Fairclough (1992) has identified hegemonic discourses related to democratization, commodification and technologization, and has demonstrated how the discourse of 'globalization' has become naturalized as the new capitalism (Fairclough undated).

Thus, discourses are loaded with values and world-views, and convey power structures. The way 'realities' are discussed and presented is always more or less consciously coloured by political or ideological stances. Spence (2007, 874), referring to Laclau (1996) discusses the relations between discourses and ideologies, stating that:

> a particular discourse shows itself as more than itself, when it projects a non-illusory image. Ideology is the denial of the illusory nature of things. When a discourse denies its own partiality and incompleteness and claims to represent the transparency or 'fullness of the community, then we can call it ideological'. (Laclau 1996, 206)

Or in simpler terms, a '[d]iscourse is seen as a means through which (and in which) ideologies are being reproduced' (Blommaert and Bulcaen 2000, 450). Ideologies are means of ordering society, and legitimizing social structures. Knowledge and professionalism are ways in which ideologies are conveyed. Through knowledge systems and discourses, ideological stances and structures are naturalized, and concealed. Ideologies legitimate not only beliefs and values, but also 'facts' and 'scientific truth'. There are a whole series of acknowledged scholars who emphasize such presumptions of knowledge (Schumpeter 1949; Berger and Luckman 1967; Geertz 1973; cf. Alvesson 1993).

In order to detect the ideological aspects of discourses, it has been claimed that they have to be deconstructed: 'Discourse is seen as a structuring principle of society, and thus through detailed deconstruction, we can examine the workings of power on behalf of specific interests and analyse the opportunities for resistance to it' (Ogbor 2000, 605). To deconstruct involves shaking up taken for granted or naturalized knowledge. In an analysis of the entrepreneurship discourse, Ogbor (2000, 607) uses 'the notion of deconstruction in order to denaturalize or call into question the knowledge claims of the entrepreneurial texts/discourses, and to reveal how they present as inherently neutral the ways things are always done'. This is about the same as the ambitions in this chapter concerning tourism destinations. This is done through applying a discourse analysis technique for identifying moments and nodal points, while naturalizing the processes and values (or ideologies) behind them.

Destinations as a Matter of Industrial Development

Neill Leiper (2008, 242) has raised questions about tourism as an industry. He sees tourism as comprising 'many nominally separate organisations, most privately owned, and others being governmental or communitarian bodies, all linked by cooperative networks that stretch from generating regions, through transit routes and into destination regions' (Leiper 2008, 242–243). He also states that tourism consists of 'collections of [...] firms [...] which to some degree and in various ways, practice cooperative and collaborative activities'. Leiper has further claimed that tourism cannot be called a real industry, because of its low profitability, weak organizations, low levels of professionalism, and lack of a strong research basis (Leiper, 1990a). Despite the fact that tourism has long been one of the world's biggest and fastest growing 'industries', it does not fit with ideas of a 'real industry', according to Leiper and others – it lacks the nodal points of the common industrial discourses. This is claimed despite the fact that these commercial activities give many people jobs, and serve the market. This is a negative way of defining an industry – stating its deficiencies and stamping it as a deviant case. Leiper's point of view is normative; claiming that one way of organizing is better than another. It is an example of streamlining, or naturalizing, as is the term used here, the thinking about tourism as an industry. This thinking has an ideal of the huge factory, with profitable large scale production. Following this logic, tourism could be seen merely as a 'subsistence area'. However, if this model was taken literally, the majority of the companies in the sector could be called deviant.

Within the realm of tourism industry thinking, there are several approaches to destination analyses. One that has been much quoted is Porter's (1992) cluster theory and diamond model. Here the success of an industry is seen as a result of the resources available (factors), relations with other industries, the competition condition (rivalry), demand condition (market), but it is also influenced by the role of authorities, and uncontrollable events:

> Porter's clusters are 'geographic concentrations of interconnected companies, specialized suppliers, service providers, firms in related industries, and associated institutions (universities, standards agencies, and trade associations) in particular fields that compete but also cooperate' ... . Such a definition could be used also to describe a destination. (Jackson and Murphy 2006, 1022)

Although intended as a model for analyses of competitiveness among nations, the model has also been applied to analyses of tourism industries in regions and destinations dominated by small and medium enterprises (SMEs) (Viken and Krogh 1995; Jørgensen et al. 2009). In these studies, the cluster effects are seen as compensating for SMEs' lack of size, and as a way of creating competitive strength. The basic idea is that the more contact between enterprises in an area, the better they perform. A low degree of interaction between companies tends to be interpreted as indicating the existence of weak clusters (Jørgensen et al. 2009). This may be a wrong conclusion. It could in fact be the discursive basis or the methodology chosen that produces such results. For instance, quantitative approaches are not suitable for detecting relationships, or influencing relations. This is demonstrated in a study in Ireland by Mottiar and Ryan (2006). They observed that despite denying collaboration with the companies around, most tourism actors cooperated and influenced each other in a number of ways, both formally and informally. And without formal collaboration, people are imitating, learning and developing, as a natural consequence of living and acting in a common neighbourhood or region. It should also be emphasized that Porter's model was not originally developed for analyses of small and medium sized enterprises (SMEs). Therefore, it may be of little value in studying tourist destinations dominated by SMEs and life style entrepreneurs. Terms such as 'competition', 'rivalry', 'demand', and 'industry' are not central to businesses based on family relations and lifestyles. However, for a long period cluster analyses of industrial regions and destinations have been conducted, for instance in Australia (Jackson and Murphy 2006), in Great Britain (Novelli et al. 2006), and in Norway (Jakobsen 1994; Viken and Krogh 1995). In Norway, the government-based Innovation Norway (a development bank, and the organization responsible for marketing Norwegian tourism abroad) has adopted the cluster model, and has financed many projects that have been aimed at creating industrial clusters. The model has defined the way destinations have been analysed and developed. This is obviously an example of the naturalization of a particular view of the industry. The Porter model has been transformed into a political or ideological mantra. Other approaches to industrial analysis include placing more focus on so called 'industrial districts'. It was through this approach that Mottiar and Ryan (2006) discovered a considerable degree of relations and reciprocal influence, despite the fact that the informants rejected cooperation. The point of departure in the approach is areas dominated by small enterprises, and how smallness can be compensated for:

An industrial district consists of a large number of small enterprises that are formally

> independent ... . The production system *per se*, however, has several features
> that could resemble a large enterprise with many separated profit centres. In
> order to obtain economies of scale, a division of labour has emerged in which
> the firms are closely linked together in horizontal, vertical and diagonal (Pyke
> 1992) webs of contracting/subcontracting and other relations. (Hjalager 1999, 4)

The growth paradigm is obvious also here; 'large enterprises', 'economies of scale', 'profit centre', and 'division of labour' are all well-known nodal points within such a paradigm. The most obvious ideological part of this account is the implicit statement of smallness being a disadvantage. However, are size and grandness the only quality or success criteria? The Mottiar and Ryan (2006) study, which presents a healthy SME-based tourism district in Ireland, shows that there are alternatives. They do in fact envision this approach as an alternative to the growth paradigm.

Approaches whereby the perspective is slightly different also exist, exemplified by those looking at destinations as learning regions (Saxena 2004) or communities of practice (Bertella 2012). They all have in common the fact that interaction is seen as a positive factor, while collaboration makes a business stronger as it enables internal specialization and constitutes a learning arena. More diffusely, collaboration produces a culture for the industry in question. In these alternative approaches the goals may differ from the growth paradigms. For instance, as Schianetz et al. (2007) have shown, creating 'learning destinations' is a path towards a more sustainable tourism. Such scholarly directions stand out as alternatives to the hegemonic 'economy of scale and scope' thinking. They at least show that in many cases there are alternative values inherent in the tourism industry – money is not the only driving force, and growth is not the only aim. However, these approaches may also be seen as implementations of neoliberal thinking (Dredge 2010) where self-criticism is praised, or as postmodern reactions to traditional industry analysis. The profit and growth paradigm is challenged by other value systems, which are partly adapted by the industries themselves. But, independent of the growth paradigm, this is of disciplinary or interdisciplinary interest in tourism studies that work for example from a social or cultural, political or geographical perspective (cf. Bramwell 2011). Among these are network and stakeholder analyses.

Destinations as Networks and Stakeholder Conglomerates

The interconnection of many functions and actors within a destination has been studied in terms of network and stakeholder conglomerates (cf. Dredge 2006; Scott et al. 2008). Pavlovich (2003) claims that tourism is a relational industry where destinations are formed by communities and groupings of organizations. Dredge (2006) reviews the network studies within the field of tourism, and identifies two major types: pure business networks and policy oriented private-public networks. She also emphasizes that network theory is a strong element in the definition of tourism as an industry by Smith (1988) and Leiper (1990a; 2008). Tinsley and Lynch emphasize the process side of networks, while networking is defined as 'cultural patterns of behaviour whose functions serve a mix of exchange, communication and social purposes'. (Leiper 2008, 18). Seen this way, networks do not only have industrial purposes, and the resources of a network can also be found outside industrial firms. Networks tend to stretch into other networks. Thus networks tie a destination both to the communities they represent and to the outer world. Often tourism network accounts have industrial or regional development, or governance of tourism, as their major analytical perspective. However, there is a tendency to place such analyses within business or growth frames. Networks are seen as a way of creating competitive advantages (Pavlovich 2003, 203) and more effective tourism management (Beaumont and Dredge 2010), suitable for market based and neoliberal policy making (ibid.). The flavour of the market economy terminology is strong frame, but it does not entirely dominate network approaches.

The stakeholder perspective opens up a new way of thinking about business, and is also a framework used for analysing industries from non-industrial perspectives. Therefore, one can also see a destination as a network of interests or stakeholders, and as including interactions between local and nonlocal actors (cf. Tinsley and Lynch 2008). Many have analysed destinations from this perspective (cf. Bærenholdt et al. 2004; Dredge 2006a, 2006b; Scott et al. 2008). Bærenholdt et al. (2004, 25) claim that 'various networking practices govern the practices of tourists, tourist businesses, tourist organisations, local authorities and other local people'. According to this view, the dynamic of tourism development lies in the tension between stakeholders. As Freeman states, '[a] stakeholder in an organization is (by definition) any group or individual who can affect or is affected by the achievement of the organization's objectives' (1984, 46). Common locations and complementary assets among stakeholders constitute a good platform for collaboration and viable networks (Tinsley and Lynch 2008). One argument for applying this model is that not only shareholders have interests in such an enterprise. Thus, it is a model where one may include whoever has a stake. In the tourism literature several discuss the stakeholder approach, among them Sautter and Leisen (1999), and Sheehan and Ritchie (2005). Sheehan and Richie have a narrow perspective, and the model is recommended as a way of detecting challenges and threats to the company. The stakeholders are seen as external actors that could influence the company, that should be met by 'appropriate management strategies' (Sheehan and Ritchie 2005, 715). The approach contrasts significantly with the principles emphasized by Sautten and Leiser (1999, 314): 'Each stakeholder group, it is further reasoned, has a *right* to be treated as an end in itself, and not as means to some other end ... ' Sheenan and Ritchie (2005) also conduct an analysis of the most important DMO (destination management organization) stakeholder groups on Hawaii; 63% of the respondents mentioned hotel and restaurants, 60% city/local government, 30% regional/county governments, 20% attractions/ attraction associations, 20% state/provincial tourism departments (Sheehan and Ritchie 2005, 728). Local people, NGOs, special interest, or culture or ethnic groups did not turn out to be significant stakeholders. This reflects the way the study was done (ibid., 720). However, it corresponds with most business oriented paradigms. Neither the local community nor the environment are reckoned to be stakeholders. The nodal points in a business oriented tourism stakeholder analysis are 'DMOs', 'the hotel and restaurant sector', and 'the authorities'. This type of 'evidence' naturalizes the business-oriented perception of the stakeholder term, at the best, towards an extended shareholder model. Others have claimed that the definition of the stakeholder map is an ethical issue (Jones et al. 2007), as is destination development (Saarinen 2004).

The network and stakeholder approaches are examples of business thinking being influenced by many disciplines and philosophies. However, alternative views do not necessary change the power situation. Network or stakeholder groups tend to operate within given power structures. Stakeholders are identified, but are they heard and followed? The stakeholder approaches can be criticized for hiding the real power structures, and in fact legitimating them – naturalizing a system where people are seen but often not heard. And in the business world, it is still the shareholders that make formal decisions, rather than the stakeholders.

Another term that is related to those discussed here is governance. According to neoliberal thinking, governance exists in the relationships between the private sector and the authorities. It is a network-based governing model. Governance is supporting neoliberalism, and represents a private-public mixed governing model, but has the character of empty terms or signifiers concerning power relations, and thus tend to hide power relations. Governance is a nodal point that seems to legitimize many projects by involving many stakeholders, giving an impression of participating and influence. Others will say that it is a manipulative model, a relational decision model that legitimates results that are negative for many stakeholders.

Destination Development as a Strategic Matter

Destination development can be an intentional and strategic matter. Strategies are based on analyses and negotiations. As Flagestad and Hope (2001) see it, it may be possible to develop a destination towards a corporation, as something that can be managed strategically (ibid., 451). The corporate model is most applicable in resorts with clear physical boundaries and a concentrated ownership, based on an ideology praising effectiveness and rationality, and a destination ideal shared by many. It is mostly found in destinations which are organized as resorts. There is a fundamental 'problem' with the community model, as they see it: the fact that most destinations are communities or regions that are not organized as business units. Destinations as communities consist of 'specialised individual independent business units ... where no unit has any dominant administrative power or dominant ownership ... ' (ibid., 452). The community model does not mean that there is total lack of strategic management; often there are DMOs and governmental organizations which take care of tourism marketing and destination management. However, this model is dominant among destinations, as most places are not constructed or merely managed as tourism destinations.

Haugland et al. (2011, 270) take as their point of departure that destinations are normally communities or regions, without a corporate structure, arguing for 'integrated multilevel destination strategies', focusing on destination capabilities, coordination at a destination level, and inter-destination bridge ties. To turn out as a strategic unit, the actors within a destination must be able to agree upon a common image and branding policy, and to utilize distributed resources and competencies in this work. This may be difficult, since the destination identity often is based on resources outside the single company. Haugland et al. (2011, 274) claim that individual actors therefore have to cooperate. And as they see it, there are four inter-organizational strategies available: working through a DMO structure, through contractual agreements, through common ownership, and to make ties to other destinations as such relations tend to stimulate both innovations and imitation. This is a strategy that reflects both the growth paradigm and neoliberal collaborative praise. For instance, collaboration is a key word in a development project related to a ski resort development discussed in this book (Viken in Chapter 7 of this volume). In general, collaboration is a nodal point in discussions about modern management, succeeding in the market, making profit and growing. Such aims need for collaborating strategies. To argue that a community matter, as with tourism development, should be handled according to business management principles, is a stance with severe political or normative implications, contradicting quite strongly the idea (and reality) that tourism development is a community task.

An examples of more classical strategic company development running into a need for collaboration, happened with the Norwegian *Hurtigruten* company's in its move into winter tourism (Viken 2011). The company operates 11 ships that go along the Norwegian western and northern coast with a daily departure from Bergen, and with Kirkenes on the Russian border as point of return. Until 2006–2007 the winter was defined as a low or non-existing season, much related to a public discourse about the north as cold, unpleasant and inaccessible. After a restructuring of their company, they examined their operation. It was pretty obvious that the winter was a loss, and it was decided to give the winter operations priority. The slogan 'Hunting the Light' was chosen (cf. Ekeland 2011). 26 new products were made – many of them produced ashore – and a massive marketing operation was undertaken in Norway, England, Germany, France and Italy. The turnaround was a success – the image of the Arctic winter was changed into something exiting and exotic. The reason for the success is threefold. First, a discursive change was made in how the ships were looked at, from transport to tourism. The company thinking changed from being related to discourses based on shipping to a discourse based on experience production. Secondly, a change in the way the product is defined can be identified; earlier, it was a boat trip for sightseeing while today, it includes a variety of onshore activities. The discursive shift also reflects changes in staff. Over the years, more people trained in tourism had been recruited. These people had other perspectives, other knowledge, and another way of looking at the ship. And, according to one of those involved, the major challenge was to change the view of the company's operation internally; the majority of the staff being stuck in the transport and shipping paradigm. And thirdly, the company started an extended collaboration with the tourism industry in the ports of call. These strategic changes, caused by the fact that many land based companies are sub-contractors, together with a couple of other winter tourism projects, and are about to transfer the whole region of North Norway into a winter destination. The public discourse therefore seems to be in a stage of naturalization. Winter tourism soon became an obvious opportunity in the eyes of the public. Thus, the Hurtigruten case is basically an example of a change of both what Somers (1992) calls ontological, conceptual and public narratives. These include new identity in the company, new ways of thinking business, new nodal points (from transport to tourism), new public opinions about the northern winter. But the overall goal was to increase profit, and to create growth.

Destination Development as a Symbolic Matter

As many have emphasized, most destinations have no natural borders, and often it is not clear what the destination is. Therefore, their outreach has to be defined. For instance, it is difficult to decide if North Norway constitutes one, three (related to three counties), or many destinations. The division of North Norway into three counties is central for a public discourse that orders the understanding of the area strongly. Although nothing indicates that this is particularly good strategy, the counties have also been used for the DMO structure. However, a destination company operating on one level does not hinder others on other levels. Destinations are often ordered in hierarchical structures (Lew and McKercher 2006), with the country on top subsuming regions, and regions including many place based destinations. As the borders for a destination are never finally settled, while new destination formations are continually negotiated, and all destinations are in fact vulnerable to discursive changes in the public (see Saarinen in Chapter 3 of this volume). A relatively new public discourse concerns place branding. Places and regions are currently seen as competitors. There are basically two groups of people devoted to the branding of places or regions. One group is tourism marketers (and DMOs) and other industries and another is local or regional authorities that try to catch the attention of investors, employees and movers (Paasi 2001; Harvey 2006). This is a relatively new way of thinking, which nobody previously saw as a matter for municipalities. As mentioned, it is part of the neoliberal ideology for towns and authorities to compete and behave according market principles (Dredge 2006).

According to Morgan et al. (2003) all destinations today can offer attractions, services and facilities, and therefore they have to differentiate and compete by portraying a unique identity. They maintain that in the future the battle for customers is being fought over hearts and minds (ibid., 286). The easy recognizable nodal points in this thinking are image and uniqueness. The aims of the branding efforts are to create strong images. Images are significant factors in the customers' choice of destinations (Lee et al. 2002). Thus, branding is necessary to identify the product in question through differentiating it from those of its competitors, and communicating a message. There are, of course, other strategies, such as positioning in relation to other destinations, or copying them. As several have shown, branding is strongly tied to a choice of logos, slogans and graphic designs (Pike 2005) – signs that convey the uniqueness of a destination and of tourist experiences to be expected. However, to create these elements for a destination can be difficult, due to the complexity, multi-dimensionality, and heterogeneity of the phenomenon in focus (ibid.). And likewise, with the heterogeneity of the customers, it is impossible to find one expression appealing to all. Also customer preferences are discursive matters, both ontologically, conceptually and based in grand narratives (cf. Somers 1996). Branding is therefore about how your message matches trends and fads in the market, and influences these trends. Branding success depends on the match between destination brand discourses and market discourses.

The marketing of the Hurtigruten company is an example of how the public discourse about the north has changed, also due to this company's marketing efforts. It probably also matches other market trends, related to search of otherness and nature based experiences. In North Finland there is an excellent example of a brand that has been employed since the late 1980s; the Santa Claus. According to a local legend, the *Joulupukin* (Santa Claus) originates from the region of Lapland, a legend also known from many other places. In the beginning there was local resistance to selling the region under this brand. However, during the years the brand has also been filled with content, for example with activities related to Christmas such as the reindeer, Santa Claus, and the white and dark winter, which turned out to be a commercial success. To develop Christmas related products has been essential. There should be a clear relationship between the brand and what is branded, between the discourses applied, and the material platform (cf. Fairclough 1999). The Santa Claus case also shows what has been observed in other places, that branding takes time (Morgan et al. 2003, 296). There are also examples of branding efforts that have failed, usually due to discursive antagonisms (cf. Laclau and Mouffe 1985). In the early 1990s, the DMOs of Northern Norway launched a campaign called 'the Green Arctic'. This was never taken up by the rest of the tourism industry in the region – they were not committed to this discourse.

Branding involves creating an image of a destination, or several images meant for different markets. However, such an image will not necessarily be based on branding. It can also be a result of a series of industrial, social and cultural processes. The Lofoten Islands is a brand appealing to a series of target groups; sightseers, mobile home owners, cyclers, angler and climbers. These markets will all tie different attributes to the region's name; *rorbuer* (fishermen's huts), fishermen's villages, steep mountains, narrow roads, cod, or fisheries. However, as the tourism industry exists today, branding efforts are being made in search of new markets; Christmas tourists, festival attendees and 'ski bums'. Branding seems to be a key part of modern tourism, even for places with strong, organic images. Place branding seems to be a fact, but one may certainly ask why such places, particularly big cities, try to recruit more tourists. Also, why should towns or other places be ready to accommodate a huge amount of travellers? Once again, it is easy to catch a glimpse of the growth and the economy of scale paradigm of capitalist and neoliberal society behind these efforts, in addition to the tourism mantras.

Destinations as Attraction Systems

Within geographical approaches to tourism, destinations are seen as manifestations of travel patterns, as particularly attractive as service structures catering for tourists, or as sites of evolution or development (Dredge 1999). Gunn has contributed to the understanding and conceptualization of destinations, and in particular as attraction complexes (Gunn 1972; 1993). Gunn regarded a destination as a region with a border, a structure of market access and circulation or a community with attractions, surrounded by a hinterland, and with entrances and gateways (Gunn 1993). His work was accompanied by a series of accounts, analysing destinations as assemblies or systems of attractions, facilities, and infrastructure needed for visitors to survive within and experience a geographically delimited area (Bukart and Medlik 1974; Murphy 1985; Mill and Morrison 1992). The systematic approach is not surprising, and a much used textbook by Mills and Morrison (1990) is entitled *The Tourism System.* This reflects the system's strong positions in many social sciences.

Leiper (1979) saw tourism as a geographical and social system consisting of four elements: the tourist generating region, the transit route region, the destination region, and the tourists and the tourism actors. He later developed a model of an attraction system, composed by the tourists, the nucleus (attractions) and its markers, in a similar way to how MacCannell (1976) had defined the attraction earlier. The following quotation shows how Leiper (1990) links the destination both to place and to attraction systems: 'A primary nucleus (attraction) is an attribute of a place, a potential tourist destination, which is influential in a traveler's decision about where to go' (Leiper 1990, 374). However, as he emphasizes in the same article, there is a whole series of spatial considerations to be taken into account in the creation of an attraction system; including questions about how many nuclei, or how many major (or primary) attractions a destination should contain, and what spatial patterns that are possible (hierarchy, clustered or scattered). The attraction system, as Leiper presents it, is strongly influenced by the spatial paradigms in geography and includes concepts such as 'space', 'place', 'region', and 'structure'. However, such models are also applicable for economically based destination analyses. An interesting part of Leiper's account is the dividing of attractions into the three categories of first, second, and third order attractions. First order attractions constitute the reason why a destination is chosen. Second order attractions are those that are known and included in the itinerary. Third order attractions are discovered on tour and bear the hallmarks of being memorable. Leiper's categorization of attractions not only proclaims differences, but creates a rank order, based on the ability to recruit customers. Thus the underlying paradigm is scale, not only an affirmation that attractions tend to be different and have different roles in the tours and product chains. Thus, the nodal points in Leipers's attraction theory are spatial and refer to the economic efficiency and business potential of attractions.

Focusing on the question of attractions, as Gunn (1972), MacCannell (1974), and Leiper (1990) have done, one can maintain that without attractions there would be no tourism or destinations. The focus of Leiper (1990) is spatial, also based on the growth paradigm. In the wake of this, analyses of attraction structures, effects of dominant attractions, or attraction shades, have been discussed (cf. Flogenfelt 2007); several of these being more spatially than economically oriented. The attraction analysis of MacCannell (1976) is based on Goffman's (1965) social analysis of front-stage and back-stage elements, and on authenticity, and on the process of 'sacralization' through which an attraction becomes a major social institution. MacCannell also uses semiology in his analyses. The attraction analysis is in fact one of the strongest non-business based approaches. However, the business paradigm has also intruded the cultural tourism research field. In a discussion of cultural attractions, Richards (2002, 1060) writes that 'individual attractions will need to think carefully about where to approach potential tourists with their marketing information'. This is rather typical for the genre, with the growth or market paradigm and langue being used as frames or contextualizations of non-growth approaches. This way of introducing articles is more or less naturalized, as it is in accounts founded in social sciences approaches to tourism.

Destinations as Places in Motion

The tourist has often been used as a metaphor for modern society (Urry 2000; Dann 2002) and is a creature on the move, but still occupied by place, living in a way that differs from that of daily life, and also known as stranger, pilgrim, vagabond, flaneur and player, among other names. Tourism destinations are his or her playgrounds. But these are also places where locals live their lives. Thus, a destination is a meeting place, and it is no surprise that tourism and travel also are metaphors used in place analyses, for instance by Massey (2005). To her, a place is essentially a 'throwntogethernes' of people, trajectories and events. The tourism destination mantra that many places take on, adds to this. This view reflects a recent twist in the social sciences, from focusing on place and stability to being occupied by movements and fluidity (Urry 2000; Giddens 2002; Urry 2007; Cresswell 2006; Bauman 2007). Within this realm, it has also been argued that the tourism destination as a place has lost some of its theoretical interest.

In most of the tourism literature, place is treated as a simple matter, and more or less as a location filled with history, materiality and a social life. There is a whole series of accounts within human geography which focus on places and regions as tourist destinations. Mostly the focus has been on optimal structuring of space, as in the works of Gunn (1973) and Leiper (1990), and those inspired by them, on the qualities of landscapes and nature, and on the significance of a particular location

(Müller and Jansson 2007; Hall et al. 2009). Destination as a place containing a social fabric or a community, has been less analysed, but is often emphasized or 'defined' into the destinations (Müller 2007). Frisk (1999, 9) defines a destination as 'a geographic local place or region where some sort of 'tourist society' inside the 'ordinary' community exists'. And as Tinsley and Lynch (2001, 374) claim, tourist places are also communities. Within such communities there are 'residents, services and businesses' that 'link the community together'. Andriotis (2008) uses the term 'integrated resorts', indicating that the intention to merge the tourist resort and the local community, and that this can be a positive thing both for the tourists and the local community. These accounts tend to put the community at the centre, focusing on nodal points such as 'community interests', 'local empowerment', 'identity', 'belonging', or on impacts (Mathieson and Wall 1988).

One example of an analysis where the destination term is avoided is a book by Bærenholdt et al. *Performing Tourist Places* (2004, 25). The authors apply a sandcastle metaphor in their analysis of tourist places, seeing them as something that disappears with 'the tide': 'Like sandcastles, tourist places are tangible yet fragile constructions, hybrids of mind and matter, imagination and presence'. Part of this story is that a tourist place is produced here and now – it is perceived differently by other people at other times. This fits well with Massey's place and space conception, but she also stresses the materiality of a place, although also seeing this as something on the move. Bærenholdt et al. (2004) do not give the local dimension of a place where people live much attention: 'Tourist places are produced by the cooperative efforts of territorially defined relations and by mobile interactions among tourist businesses, tourist organization and tourists across place boundaries'. (ibid., 25). Their focus is more on the staging, performances and semiotics of place, and on places as arenas for play, performance and pleasure. These concepts are the nodal points in their story. In the book, the discourse about tourists is prolonged into a ludic academic account, focusing on the fun sides of tourism, such as the beach, the photography, good memories, and the 'good old times'. The market thinking is also there: tourist places are 'produced', 'consumed', and consist of 'networks' and 'businesses'. The flavour of the marketing approach is easy to recognise.

There is one particular field within tourism studies that focuses on place as a location for communities. This is known as 'community based tourism' (CBT). Here community involvement is seen as a necessity for creating sustainable tourism (Blackstock 2005), and according to Pearce (1992) this should mean local control, consensus and an equitable flow of benefits from the tourism operations. This may be a good model or approach to tourism when there is a strong wish to preserve a community as it is. The involvement of local people is many places a naturalized public discourse, with nodal points as 'empowerment', 'involvement', participation', and 'democracy', and challenging the position and power of the growth paradigm. Normally, and particularly if growth occurs, such ideals are difficult to put into practice. For example, Blackstock (2005) poses the following question: Who is representative for the community in such questions? An even more fundamental question is whether one can control a growth development, in a world with freedom to move both for people and capital. The democratic idea of participation collides with the democracy of capital. The CBT thinking has focused on a difficult time, when places, to apply Massey's terms, are made up though 'throwntogetherness', trajectories and events; aspects of places that tourism tend to accelerate. Thus, there is a possibility that community based tourism projects will stand out as a local and political cry against globalized processes that everybody more or less, and particularly within tourism, worships.

Normative Destination Discourses

Tourism has severe societal implications. The awareness of this has created a whole series of approaches to destination analyses and destination development. This reflects the emergence of tourism research, which grew back in the 1970s, as the negative impacts of tourism began to become evident. Natural, social and cultural changes and impacts were also discussed (Mathieson and Wall 1980), along with the cultural and social transformations involved in those phenomena (Smith 1977). The major discourse was about mass tourism and its character and its impact on local environments. It was also discussed if there were limits to growth in this sector. After the Brundtland Report (1987) the discourse changed. In this report the world's environmental problems were addressed, but the agenda changed from the idea of limiting economic development to promoting sustainable development. Such development aimed at offering the next generation the same opportunities to live off natural and cultural resources as contemporary society. Growth was seen as the way out of environmental problems. The mantra was that the knowledge and technology that had created the growth and that corresponding environmental problems, should be used to solve these problems. Behind this was a discourse on sustainability, which can be traced back to eco-philosophers such as Huber (1982), and the idea of ecological modernization. Sustainability soon became a strong public and political discourse, followed by a huge amount of scholarly accounts. The idea of sustainable development also gave rise to a whole series of tourism research fields, which more or less kept to the philosophy of the Brundtland Report, where sustainable tourism (cf. Mowforth and Munt 1998; Hall and Lew 1998), ecotourism (cf. Fennell 2003), responsible tourism (cf. Harrison and Husbands 1996), and ethical tourism (cf. Fennell 2006) constituted the most significant trends. The significance of this perspective on tourism was also focused on in journals like *Journal of Sustainable Tourism* and *Journal of Ecotourism.* Most of the accounts found in these tourism books and journals concentrate on ecological modernization and sustainability, while representing what others have called soft or weak environmental approaches (Hay 2002), where ecological concern is a principle and not an imperative.

Some discursive aspects of this tradition should be mentioned. The Brundtland Report must be said to have had a strong impact on society in general. Environmental issues became a public as well as an industrial discourse. This seems to have become institutionalized, and become a more or less naturalized perspective (Eder 1992). Concerning tourism destinations, there is a long series of accounts of sustainability, ecotourism and cultural change. One of the discussions, first raised by Butler (1992), regards the question of whether the sustainability principle is applicable on a company, resort or destination level, or to tourism in general. Most environmental problems are common burdens; to have sustainable strategies in one location is of little benefit if the neighbour actors or areas do not have them. This is one of the paradoxes of the slogan 'Think Globally, Act Locally' which was launched at the Rio Summit Conference in 1992, a UN follow-up meeting to the Brundtland Report. However, this meeting and the public discourse which followed somehow turned sustainability and environmental issues into a personal problem as well as a challenge for local communities. Thus, it was expected that tourism destinations should also take on this burden. However, and despite the many positive examples that could be identified, it is rather unclear if the tourism industry as such is moving in an environmentally sustainable direction. This is, however, an ongoing discourse. One of the trends after the Rio Summit was to talk about responsible tourism while seeing all levels as responsible for what is going on in the tourism field: the tourists, the tour operators, the transport sector, the tourism companies at the destinations, and the authorities. However, tourism is related to that of moving and being an intruder in a foreign environment. Thus, the essence of the phenomenon in question is a paradox. A tourism destination is by definition a place where several sustainability questions emerge, but can never be solved. However, and this is important, it is better to have such aims, and to make such considerations, than not. Another benefit from these discourses is that industry actors and destination developers have become more reflexive about their activities.

A Critical Discourse of Destinations

There are of course a series of accounts which are critical of tourism and destination development. Some of the critics produce variants of a general critique of western society or the capitalist economy, whereby the problems of poverty, inequality and environmental degradation are related to western consumption and lifestyles. Issues about various destinations also have been analysed under headlines like 'spectacularization', 'Disneyfication', 'McDonaldization' and 'disenchantment'. Such concerns are often related to strong contemporary metanarratives such as globalization and cultural homogenization. Some of the critiques go back to Marx. For instance, the International Situationists in France during the 1960s and 1970s, and particularly Debord (1970), were preoccupied with the spectacularization of society (cf. Gardiner 2000; Best and Kellner, undated) and trends invoked by the capitalist economy that turn people into consumer dupes. In the wake of these accounts several authors have analysed tourism landscapes (Gottdiener 1997; Gotham 2005). For example Ritzer (1999; 2004) has presented a series of analyses about the disenchantment and re-enchantment of society, related to processes of rationalization, standardization, and the postmodern reaction. There are of course also other critical stances. Within tourism academia, an association has been established known as A Critical Turn in Tourism Theory. Such analyses and movements are vital to taking on a reflexive tourism destination analysis.

The fact that destinations are socio-spatial and discursive matters make destinations exposed to discursive shifts related to economic as well as socio-cultural, political, and academic spheres. One of the discourses in action since the 1980s has been about globalization, a trend to which also tourism contributes considerably. Tourism has been called a hyper-globalizer. Hjalager (2007) has shown how globalization is working within tourism in the economic sector, through tourist boards being represented all over the world, marketing collaborations, and companies operating in several countries, as well as through transnational ownership, interchange of business concepts, outsourcing, international workforces, and what she has called fragmented and new types of value chains, such as stronger ties to knowledge production. Such processes can be observed all over the world. This discussion is well situated within the growth discourse.

A proponent of the critique of globalization is Ritzer (2008), who has written about the 'globalization of nothing'. He uses tourism appearances as examples of this 'nothingness'. He sees theme parks, manufactured souvenirs, 'non-places' and standardized service personnel as aspects of modern destinations. Ritzer claims that globalization has no other content than that of strengthening the positions of those behind modern capitalism, for example those gaining from a practice according to the growth paradigm. In a similar way, Terkenli (2002) discusses globalization within the context of the cultural economy and postmodernism and argues that destination landscapes have been strongly transformed. First, he points at what he calls *enworldment*. This is related to the collapse of geographical borders and to the fusion of life worlds as well as nature and culture. Shapes, concepts, modes and basic services are produced in similar ways all over the world, to please modern consumers. Terkenli (2002, 237) explains that this is a mixture of 'old and new' and of 'familiar and different'. The process of enworldment is often linked to the process of *unworldment*, such as the processes that detach people from place, as expressed within discussions of placelessness and non-places. As places are social products, and people are less attached to places, the distinctness of places tends to be feebler, he argues. This point is also stressed by others (cf. Urry 1990). The world is presented in a recognizable and predictable fashion, enabling the tourist to move about without undue concern or shock. When places are invaded of tourists, they are produced by the motions of people and other flow elements. Thus, in many places there is not much signification of local roots and culture, and as destinations they compete in their ability to deviate from the mainstream trends of the enworldment processes. According to Terkenli, the way destinations answer is by *deworldment*, through 'the creation of fictious, commercialized, ephemeral, disposable, staged, 'inauthentic' landscapes and worlds' (ibid., 242). This is the tourism production system with its theme parks and fantasy worlds. Terkenli maintains that this tends to be a masculine world. Also, there is a process of *transworldment,* as a result of which world is no longer seen as being material, but as virtually and electronically mediated and consumed. Thus the consumer becomes unable to recognize the difference between the real and virtual. The virtual tends to be perceived of being authentic, and the real as false. How shall destinations and consumers navigate in this landscape? How to come out of these processes with genuine places and attraction systems and images?

Saarinen (2004) is less pessimistic. He sees a duality in the globalizing and homogenizing forces, which also provide room for local culture, and local manifestations of global trends. Destinations are still linked to local materialities of place and practices from the past. One may also say that globalization has made possible new ways of enriching local cultures. And further, globalization is a delocalization process creating greater reach and accessibility.

Based on this discussion – a situation where all destinations more or less offer the same facilities while being exposed to the same trends – it is also reasonable to ask if 'destination' is the right term, or at least the right unit for marketing efforts. There is a tendency towards specialized consumption (see Viken and Granås in Chapter 1 of this volume). Tourists do not search for destinations as areas with varieties of offers, but as places that provide the particular type of experiences being searched for. And if and when a destination specializes in a particular activity, such as diving, hiking, or biking, there is a risk that the destination as such will become less important – it is the specialized products and the infrastructure that counts. This may represent a dis-embedment of destinations, and support for the 'marketization' and the growth paradigm. The discourses that have evolved during recent decades about tourist experiences, active holidays, sports and self-realization make the question of 'where' less important. There are tendencies in the market which minimize the role of the general DMOs. In some towns different organizations take on the role of DMOs, and a split between tourism and convention bureaus is quite common. If it is not the place or regions that attracts, another platform could be chosen.

There are also reasons to pose questions about the validity and rationality of such discourses. Are globalization and cultural homogenization ideological projects trying to convince people to give up their local production and production traditions and culture, paving way for global corporations or other companies, like national and international chains, that play the same game? And what roles do such discourses have? Are critical voices only another part of the neoliberal and postmodern project?

Conclusion: How to Contest the Growth Paradigm within Destination Development?

This chapter is a review of some of the more commonly used approaches in the analysis of tourism destinations. The focus has been on the discourses and ideologies that are conveyed, with particular regard to the growth paradigm. This paradigm has a naturalized place in the economically based discourses, but seems also to have invaded many social scientific destination accounts. There are, however, alternative approaches. Although not all of these are discussed here, there are anthropological studies of destinations that have hardly been influenced by the growth paradigm. The discussions of sustainable, responsible or ethical tourism are also significant, but tend to consist of attacks on the industry's ignorance of such perspectives and values. The most important argument that can be put forward about such an analysis is that the understanding of advantages of the growth paradigm – growth, big units, international capital, and so on – seems to have a naturalized position; more or less threated as a natural law and a truth. Such normative or ideological stances obviously contradict the realities in the field, in which SMEs, micro- or nano-sized companies do in fact dominate. As this chapter shows, the reality does not fit the theory, but according to the viewpoint taking growth and the economy of scale for granted, it is the 'reality' that is wrong.

There is no doubt that the industrial and business approaches to destinations and destination development are dominant, but are also a necessity. However, as is also argued here, these discourses are not neutral and fit big companies, and probably urban districts more than the rural SME sector. Thus, the prevailing destination thinking is not necessarily the best mantra for all, and certainly not for actors outside the business world. Destinations are places containing local communities that should be analysed and understood as such, within the context of tourism. One of the questions to be raised in the wake of this, is to what extent scholars and actors in the field in producing and applying these models are aware of their discursive stances. Despite all the relativism and deconstructionism, epistemological perspectives, discourses and ideological stances tend not to be problematized. It is not only the researchers that are 'hidden' (Tribe 2005), in many of the prevailing texts. As argued here, the growth paradigm is also often hidden, along with the discourses and ideologies that lie behind it.

References

Abram, S., Waldren, J. and Macleod, D. (eds) (1997), *Tourists and Tourism: Identifying with People and Places* (Oxford: Berg).

Aitchison, C. (2000), 'Poststructural Feminist Theories of Representing Others: A Response to the 'Crisis' in Leisure Studies' Discourse', *Leisure Studies* 19:3, 127–144.

Alvesson, M. (1993), 'Organizations as Rhetoric: Knowledge-Intensive Firms and the Struggle with Ambiguity', *Journal of Management Studies* 30:6, 997–1015.

Andriotis, K. (2008), 'Integrated Resort Development: The Case of Cavo Sidero, Crete', *Journal of Sustainable Tourism* 16:4, 428–444.

Bærenholdt, J.O., Haldrup, M., Larsen, J, and Urry, J. (2004), *Performing Tourist Places* (Aldershot: Ashgate).

Bauman, Z. (2007), *Liquid Times: Living in an Age of Uncertainty* (Cambridge: Polity).

Beaumont, N. and Dredge, D. (2010), 'Local Tourism Governance: A Comparison of Three Network Approaches', Journal of Sustainable Tourism 18:1, 7–28.

Berger, P.L. and Luckmann, T. (1967), *The Social Construction of Reality* (New York: Doubleday).

Bertella, G. (2011), 'Communities of Practice in Tourism: Working and Learning Together. An Illustrative Case Study from Northern Norway', *Tourism Planning and Development* 8:4, 381–397.

Best, S. and Kellner, D. (undated), 'Debord, Cybersitations and the Interactive Spectacle, *Illuminations*. The Critical Theory Website'. http://www.uta.edu/huma/illuminations/ (February 24, 2009).

Blackstock, K. (2005), 'A Critical Look at Community Based Tourism', *Community Development Journal* 40:1, 39–49.

Blommaert, J. and Bulcaen, C. (2000), 'Critical Discourse Analysis', *Annual Review of Anthropology* 29, 447–66.

Bohlin, M. and Elbe, J. (eds) (2007), *Utveckla turistdestinationer. Ett svenskt perspektiv* (Uppsala: Uppsala Publishing House).

Bornhorst, T., Ritchie, J.R.B. and Sheehan, L. (2010), 'Determinants of Tourism Success for DMOs and Destinations: An Empirical Examination of Stakeholder' Perspectives', *Tourism Management* 31:5, 572–589.

Bourdieu, P. (1967), 'Systems of Education and Systems of Thought', in Young (ed.).

Bourdieu, P. (1986), 'The Forms of Capital', in Richardson (ed.).

Bourdieu, P. (2004), *Science of Science and Reflexivity* (Palo Alto, CA: Stanford University Press).

Bramwell, B. (2011), 'Governance, the State and Sustainable Tourism: A Political Economy Approach', *Journal of Sustainable Tourism* 19:4/5, 459–477.

Bramwell, B. and Lane, B. (2000), 'Collaboration and Partnerships in Tourism Planning', in Bramwell and Lane (eds).

Bramwell, B. and Lane, B. (eds) (2000), *Tourism Collaboration and Partnerships. Politics, Practice and Sustainability* (Clevedon: Channel View Publications).

Burkart, A.J., and Medlik, S. (1974), *Tourism. Past, Present and Future* (London: Heinemann).

Butler, R. (1980), 'The Concept of a Tourist Area Cycle of Evolution: Implications for Management of Resources', *Canadian Geographer* 24:1, 5–12.

Butler, R.W. (1992), 'Alternative Tourism: The Thin End of the Wedge', in Smith and Eadington (eds).

Carlsen, J. (1999), 'A System Approach to Island Tourism Destination Management', *Systems Research and Behavioral Science* 16:4, 321–327.

Cresswell, T. (1996), *On the Move: Mobility in the Modern Western World* (New York: Routledge).

Dann, G. (2002), 'The Tourist as a Metaphor of the Social World', in Dann (ed.).

Dann, G. (ed.) (2002), *The Tourist as a Metaphor of the Social World* (Wallingford: Cabi).

Davidson, P. and Honig, B. (2003), 'The Role of Social and Human Capital among Nascent Entrepreneurs', *Journal of Business Venturing* 18:3, 301–331.

Debord, G. (1987), *Society of the Spectacle* (London: Rebel Press).

Derrida. J. (1978), *Writing and Difference* (London: Routledge and Kegan Paul).

Dredge, D. (1999), 'Destination Place Planning and Design', *Annals of Tourism Research* 26:4, 272–791.

Dredge, D. (2006), 'Policy Networks and the Local Organisation of Tourism', *Tourism Management* 27:2, 269–280.

Eder, K. (1996), 'The Institutionalisation of Environmentalism: Ecological Discourse and Second Transformation of the Public Sphere', in Lash et al. (eds).

Fairclough, N. (1989), *Language and Power* (London: Longman).

—— (1992), *Discourse and Social Change* (Cambridge: Polity Press).

Fennell, D. (2003), *Ecotourism: An Introduction* (2nd edn) (London: Routledge).

—— (2006), *Tourism Ethics* (Clevedon: Channel View Publications).

Flagestad, A. and Hope, C.A. (2001), 'Strategic Success in Winter Sports Destinations: A Sustainable Value Creation Perspective', *Tourism Management* 22:5, 445–461.

Flognfeldt, T. (2007), 'Developing Tourism Products in the Primary Attraction Shadow', *Tourism Culture and Communication* 7:2, 133–145.

Framke, W. (2002), 'The Destination as a Concept: A Discussion of the Business-Related Perspective versus the Socio-Cultural Approach in Tourism Theory', *Scandinavian Journal of Hospitality and Tourism* 2:2, 92–107.

Freeman, R.E. (1984), *Strategic Management: A Stakeholder Approach* (Boston: Pitman).

Fries, C.J. (2009), 'Bourdieu's Reflexive Sociology as a Theoretical Basis for Mixed Methods Research', *Journal of Mixed Methods Research* 3:4, 326–328.

Frisk, L. (1999), *Separate Worlds: Attitudes and Values towards Tourism Development and Cooperation among Public Organisations and Private Enterprises in Northern Sweden* (Östersund: Proceedings of Forskarforum: Local och Regional Utveckling).

Gardiner, M.E. (2000), *Critiques of Everyday Life* (London: Routledge).

Geertz, C. (1973), *The Interpretation of Cultures* (New York: Basic Books).

Giddens, A. (2002), *Runaway World: How Globalisation is Reshaping our Lives* (London: Profile Books).

Gotham, K.F. (2005), 'Theorizing Urban Spectacles. Festivals, Tourism and the Transformation of Urban Space', *City* 9:2, 225–246.

Gottdiener, M. (1995), *Postmodern Semotics. Material Culture and the Forms of Postmodern Life* (Oxford: Blackwell).

Gramsci, A. (1971), *Selection from the Prison Notebooks* (New York: International Publishers).

Gunn, C. (1972), *Vacationscape: Designing Tourist Regions* (Austin: University of Texas).

Gunn, C. (1979), *Tourism Planning* (New York: Crane Russack).

Hajer, M. (1995), *The Politics of Environmental Discourse* (Oxford: Oxford University Press).

Haldrup, M. (2001), *Tourists in Time and Space – Mobility Tactics and the Making of Touristspaces. Draft.* (Roskilde: Roskilde University).

Hall, C.M. and Lew, A.A. (eds) (1998), *Sustainable Tourism Development: Geographical Perspectives* (Harlow: Addison Wesley Longman).

Hamilton C. (2003) *Growth Fetish* (London: Pluto Press).

Harrison, L. C. and Husbands, W. (1996), *Practicing Responsible Tourism: International Case Studies in Tourism Planning, Policy and Development* (New York: John Wiley and Sons).

Haugland, S.A., Ness, H., Grønseth, B.-O. and Aarstad, J. (2011), 'Development of Tourism Destinations. An Integrated Multilevel Perspective', *Annals of Tourism Research* 38:1, 268–290.

Hay, P. (2002), *A Companion to Environmental Thought* (Edinburgh: Edinburgh University Press).

Herman, E. and O'Sullivan, G. (1989 [1988]), *The 'Terrorism' Industry: The Experts and Institutions that Shape Our View of Terror* (New York: Pantheon Books).

Higgins-Desbiolles, F. (2006), 'More than an "Industry": The Forgotten Power of Tourism as a Social Force', *Tourism Management* 27:6, 1192–1208.

Hjalager, A.-M. (1997), 'Innovation Patterns in Sustainable Tourism. An Analytical Typology', *Tourism Management* 18:1, 35–41.

—— (1997), *Tourism Destinations and the Concept of Industrial Districts.* Paper for ERSA conference, Dublin, August 1999 (Århus: Science Park).

—— (2007), 'Stages in the Economic Globalization of Tourism', *Annals of Tourism Research* 34:2, 437–457.

Huber, J. (1982), *Die verlorene Unschuld der Ökologie. Neue Technologien und superindustrielle Entwicklung* (Frankfurt: Fisher).

Ioannides, D. and Debbage, K.G. (eds) (1998), *The Economic Geography of the Tourism Industry* (London: Routledge).

Jackson, J., and Murphy, P. (2006), 'Clusters in Regional Tourism: An Australian Case', *Annals of Tourism Research* 33:4, 1018–1035.

Jamal, T., and Getz, D. (1995), 'Collaboration Theory and Community Tourism Planning', *Annals of Tourism Research* 22:1, 186–204.

Jørgensen, E., Abelsen, B. and Korsnes, K. (2008), *Klynge eller praksisfellesskap? En studie av samhandling mellom reiselivsbedrifter i Finnmark.* Rapport Norut Finnmark 11 (Alta: Norut Alta).

Knoke, D. and Kuklinski, J.H. (1991), 'Network Analysis. Basic Concepts', in Thompson et al. (eds).

Kögler, H.H. (1997), 'Reconceptualizing Reflexive Sociology: A Reply', *Social Epistemology* 11:2, 225–250.

Laclau, E. and Mouffe, C. (1985), *Hegemony and Socialist Strategy* (London: Verso).

Laclau, E. (1996), 'The Death and Resurrection of the Theory of Ideology', *Journal of Political Ideologies* 1:3, 201–20.

Larner, W. and Craig, D. (2005), 'After Neoliberalism? Community Activism and Local Partnerships in Aotearoa New Zealand', *Antipode* 37:3, 402–424.

Lash, S. et al. (eds) (1996), *Risk, Environment and Modernity* (London: Sage).

Lee, L.G., O'Leary, J.T. and Hong, G.S. (2002), 'Visiting Propensity Predicted by Destination Image', *International Journal of Hospitality and Tourism Administration* 3:2, 63–92.

Leiper, N. (1979), 'The Framework of Tourism: Towards Definitions of Tourism, Tourists and the Tourism Industry', *Annals of Tourism Research* 6:4, 390–407.

—— (1990a), 'The Business of Tourism and the Partial-Indusrialization of Tourism Systems: A Management Perspective', in Leiper (ed.).

——(1990b),'Tourist AttractionSystems', *AnnalsofTourismResearch*17:3,367–384.

—— (2008), 'Why 'the Tourism Industry' is Misleading as a Generic Expression: The Case for the Plural Variation, 'Tourism Industries', *Tourism Management* 29:2, 237–251.

Leiper, N. (ed.) (1990a) *Tourism Systems: An Interdisciplinary Perspective* (Palmerston: Massey University).

Lew, A. and McKercher, B. (2006), 'Modeling Tourist Moments. A Local Destination Analysis', *Annals of Tourism Research* 33:2, 403–423.

MacCannell, D. (1999 [1976]), *The Tourist. A New Theory of the Leisure Class* (Berkeley: University of California Press).

MacCannell, D. (1992), *Empty Meeting Grounds. Class* (New York: Routledge).

Masey, D. (2005), *For Space* (London: Sage).

Mathieson, A. and Wall, G. (1982), *Tourism: Economic, Physical, and Social Impacts* (New York: Longman House).

Meethan, K. (2001), *Tourism in Global Society. Place, Culture, Consumption* (Basingstoke: Palgrave).

Mill, R. C. and Morrison, A. M. (1998), *The Tourism System: An Introductory Text* (3rd edn) (Dubuque, IA: Kendall/Hunt Publishing Company).

Morgan, N., Pritchard, A. and Piggott, R. (2003), 'Destination Branding and the Role of Stakeholders: The Case of New Zealand', *Journal of Vacation Marketing* 9:3, 285–299.

Morrison, A. (2006), 'A Contextualisation of Entrepreneurship', *International Journal of Entrepreneurial Behaviour and Research* 12:4, 192–209.

Mottiar, Z. and Ryan, T. (2012)', The Role of SMEs in Tourism Development: An Industrial District Approach Applied to Killarney, Ireland', in Thomas and Augustyn (eds).

Mowforth, M. and Munt, I. (1998), *Tourism andSustainability* (London:Routledge).

Müller, D. (2007), 'Planering för turistdestinasjoner', in Bohlin and Elbe (eds).

Murphy, P. (1985), *Tourism: A Community Approach* (London: Routledge).

Novelli, M., Schmitz, B., and Spencer T. (2006), 'Networks, Clusters and Innovation in Tourism: A UK Experience', *Tourism Management* 27:6, 1141–1152.

Ogbor, J. (2000), 'Mythicizing and Reification in Entrepreneurial Discourse: Ideology-Critique of Entrepreneurial Studies', *Journal of Management Studies* 37:5, 605–635.

Pavlovich, K. (2003), 'The Evolution and Transformation of a Tourism Destination Network: The Waitomo Caves, New Zealand', *Tourism Management* 24:2, 203–216.

Phillips, L. and Jørgensen, M.W. (2002), *Discourse Analysis as Theory and Method* (London: Sage).

Pearce, D. (1992), 'Alternative Tourism: Concepts, Classifications and Questions', in Smith and Eadington (eds).

Pickard, D. (2011), *Tourism, Magic and Modernity. Cultivating the Human Garden* (London: Berghahn).

Pike. S. (2005), 'Tourism Destination Branding Complexity', *Journal of Product and Brand Management* 14:4, 258–259.

Porter, Michael E. (1990), *The Competitive Advantage of Nations* (New York: The Free Press).

Pyke, F. (1992), *Industrial Development through Small-Firm Cooperation, Theory and Practice* (Geneva: ILO).

Reve, T. and Jakobsen, E.W. (2001), *Et verdiskapende Norge* (Oslo: Universitetsforlaget).

Richards, G. (2002), 'Tourism Attraction Systems: Exploring Cultural Behavior', *Annals of Tourism Research* 29:4, 1048–1064.

Richardson, J. (ed.) (1986), *Handbook of Theory and Research for the Sociology of Education* (Westport, CT: Greenwood Press).

Ritzer, G. (1999), *Enchanting a Disenchanted World: Revolutionizing the Means of Consumption* (Thousand Oaks: Pine Forge).

Ritzer, G. (2004), *The Globalization of Nothing* (Thousand Oaks, CA: Pine Forge Press).

Sautter, E.T. and Leisen, B. (1999), 'Managing Stakeholders: A Tourism Planning Model', *Annals of Tourism Research* 26:2, 312–28.

Saarinen, J. (2004), 'Destinations in Change. The Transformation Process of Tourist Destinations', *Tourist Studies* 4:2, 161–179.

Saxena, G. (2004), 'Relationships, Networks and the Learning Regions: Case Evidence from the Peak District National Park', *Tourism Management* 26:2, 277–289.

Schianetz K, Kavanagh L, Lockington D. (2007), 'The Learning Tourism Destination: The Potential of a Learning Organisation Approach for Improving the Sustainability of Tourism Destinations', *Tourism Management* 28:6, 1485–1496.

Schumpeter, J.A. (1934), *The Theory of Economic Development* (reprint edn 1959) (Cambridge, Mass.: Harvard University Press).

Schumpeter, J. A. (1949), 'Science and Ideology', *American Economic Review* 39:2, 345–59.

Scott, N., Cooper, C. and Baggio, R. (2008), 'Destination Networks: Four Australian Cases', Annals of Tourism Research 35:1, 169–188.

Smith, V.L. (ed.) (1977), *Hosts and Guests: The Anthropology of Tourism* (Philadelphia: University of Pennsylvania Press).

Smith, V.L. and Eadington, W. (eds) (1992), *Tourism Alternatives: Potentials and Problems in the Development of Tourism* (Chichester: Wiley).

Smith, S.L.J. (1998), 'Tourism as an Industry', in Ioannides and Debbage (eds).

Somers, M.R. (1994), 'The Narrative Constitution of Identity: A Relational and Network Approach', *Theory and Society* 23:5, 605–649.

Spence, C. (2007), 'Social and Environmental Reporting and Hegemonic Discourse', *Accounting, Auditing and Accountability Journal* 20:6, 855–882.

Terkenli, F. S. (2002), 'Landscapes of Tourism: Toward a Global Cultural Economy of Space?' *Tourism Geographies* 4:3, 227–254.

Thomas, R. and Augustyn, M. (eds) (2012), *Tourism in the New Europe: Perspectives on SME Policies and Practices* (forthcoming) (London: Elsevier).

Thompson, G. et al. (eds) (1991), *Markets, Hierarchies and Networks: The Co-ordination of Social Life* (London: Sage).

Tinseley, R. and Lynch, P. (2001), 'Small Tourism Business Networks and Destination Development', *International Journal of Hospitality Management* 20:4, 367–378.

—— (2008), 'Differentiation and Tourism Destination Development: Small Business Success in Close-Knit Community', *Tourism and Hospitality Research* 8:3, 161–177.

Tribe, J. (2005), 'New Tourism Research', *Tourism Recreation Research* 30:2, 5–8.

Urry, J. (1990), *The Tourist Gaze* (London: Sage).

Urry, J. (2000), *Sociology Beyond Societies: Mobilities for the Twenty-First Century* (London: Routledge).

Urry, J. (2007), *Mobilities* (Cambridge: Polity).

Van Dijk, F.A. (1993), 'Principles of Critical Discourse Analysis', *Discourse and Society* 4:2,249–283.

Vázquez, G. P. (2006), 'The Recycling of Local Discourses in the Institutional Talk: Naturalization Strategies, Interactional Control, and Public Local Identities', *Estudios de Sociolingüística* 7:1, 55–82.

Viken, A. and Krogh, L. (1995), *Reiselivsnæringen.* (Et konkurransedyktig Nord-Norge). Rapport nr. 2. (Alta: Finnmarksforskning).

Viken, A. (2011), 'Hurtigruten på vinterkjøl', chronicle, *Nordlys* 3.3.2011.

Wall, G. (1996), 'Integrating Integrated Resorts', *Annals of Tourism Research* 23:3, 713–717.

Young, M.F.D. (ed.), *Knowledge and Control* (London: Collier-Macmillan).

Chapter 3
Transforming Destinations: A Discursive Approach to Tourist Destinations and Development

Jarkko Saarinen

Introduction

Change is a said to be a constant feature in tourism. As stated by Richard Butler (1980, 5) in his seminal work on the evolution cycle of tourism, 'there can be little doubt that tourist areas are dynamic, that they evolve and change over time'. The aspect of change is not only related to the patterns and flows of supply and demand in tourism but also the very spaces of tourism: the tourist destinations. In general, destinations are sites connecting the activities and processes of consumption and production in the tourism system (see Leiper 1990; Hall 1997) and during the past two decades the discussion regarding tourist destinations and their changes has increasingly stressed the meaning and role of space and spatiality in development (see Ashworth and Dietvorst 1995; Meethan 1996, 2001; Saarinen, 1998; Gordon and Goodall 2000; Franklin and Crang 2001; Ateljevic and Doorne 2002; Gale and Botterill 2005; Saarinen and Kask 2008). As argued by John Urry (1995, 1): 'almost all the major social and cultural theories bear upon the explanation of place', and this notion is very relevant for tourist destinations, their change and nature.

Contemporary tourism and tourist destinations are increasingly transformed by socio-economic and political forces, systems and relations in a local-global nexus (see Aitchison et al. 2000; Saarinen 2004). A non-locally driven development is not a new issue in tourism, but tourist destinations are currently being transformed on a non-local basis and more rapidly than ever before (see Terkenli 2002). According to Williams and Baláž (2002, 39), the most obvious manifestation of the deepening of relations and processes across space and time is the massive growth in global tourism. This is evident in many places in the world, including the Nordic region and its peripheries, where new evolving destinations and tourist activities have become increasingly characteristic features of the changing localities and societies (see Viken 1995; Hall et al. 2009; Hall and Saarinen 2010). While globalization is not a new phenomenon in tourism (see Featherstone and Lash 1995), the deepening processes involving increasing changes, competition, enclavization and flows of capital, ideas and materials etc. have created a need for analysing destinations and their transformation processes in a way that considers tourism spaces in relational settings connecting different scales and regional units.

This chapter discusses the conceptual idea of tourist destinations and their transformation processes as a set of discursive practices. On one hand, these practices are seen as operating in a certain socio-spatial context and on the other hand in a global-local nexus interconnecting distant places and creating deepening dependency between different regions and scales. The chapter is based on previous papers (see Saarinen 1998; 2004; Saarinen and Kask 2008) and aims to further develop and re-contextualize the framework of the transformation process of tourist destinations.

Tourist Destinations

Focus on tourist destination change and development has been a core area in tourism studies (see Gilbert 1939, 1949; Christaller 1963; Miossec 1967; Butler 1980; Shields 1991; Meethan 1996; Johnston 2001; Agarwal 2002). Similarly, the connections between tourism industry and regional development, i.e., relationships between tourism growth and surrounding socio-spatial structures and processes, have interested tourism researchers for a relatively long period of time. Although studies focusing on tourist destinations have a long history, Michael Haywood (1986) has stated that the past research has given relatively little attention to the conceptual nature and identification of the unit of analysis, the tourist destination (see also Lew 1987; Cooper 1989; Framke 2002; Saarinen 2004).

While Haywood's (1986) statement was published almost three decades ago, it is still a relevant evaluation. Indeed, destination has remained as a problematic concept which can be used in different ways and in a varying range of spatial scales. In some contexts it refers to continents, other world regions or states, while also provinces, municipalities and other local administrative units, single resorts or even attraction sites can be termed destinations in tourism research, development policies and strategies. In tourism policy and development discussions, different regional units can also be mixed and in conflict with other existing regional classifications and scales creating a complex landscape of regions and boundaries to interpret. In the UNWTO's regional tourism statistics, for example, Egypt as a country (destination) is not part of the continent of Africa or the North Africa region but the Middle East region.

Connecting the idea of tourist destination to a certain administrative unit is widely used strategy in research and regional development analysis and policy-making. It is often a practical approach, although usually taken as a granted connection – or the connection is justified based on the reasons referring to the nature of statistical system, for example, i.e. that tourism and other socio-economic data in use have been collected based on administrative units, such as municipalities or the NUTS (nomenclature of territorial units for statistics) regions in the European Union (see Saarinen 2003). The NUTS is a hierarchical system for dividing up the economic territory of the EU for the purposes of the collection, development and harmonization of EU regional statistics and socio-economic analyses of the regions (see Eurostat 2013), including the tourism industry. The statistics based on the NUTS system are then used for regional development evaluations, policy-making and decisions. Therefore, the conceptual understanding of region can have a great significance in practice.

An administrative basis to define tourist destinations can relate to the nature of tourism development agencies and actions: e.g. regional authorities investing in and promoting tourism development are based in and in many cases also limited to administrative spatial units, such as provinces or the mentioned EU NUTS regions. Thus, the development actions operate in bounded spaces that are limited within the power of authorities and their territorial space. However, the administrative perspective tends to approach tourist destinations and tourism from a technical and spatially static or homogenous viewpoint. In the end, however, tourist destinations are not necessarily territorially limited bounded entities but parts of relational processes where tourist activities and tourism development do not necessarily follow the boundaries of administrative units but evolve over the limits of territories which may be insignificant for tourists, the industry or the regional identity of destinations (Saarinen 1998, 2004). In addition, only certain small parts of administrative regions may be meaningful in a touristic sense. Generalization of those 'tourist hot spots' to a wider administrative regional unit may be misleading in development work and strategies. These kinds of tourist hot spots are rather typical cases in tourism development in peripheries where the spaces of tourism can evolve as enclaves with limited connections to surrounding (administrative) regional socio-economic structures and processes (see Britton 1979; 1991; Saarinen 2004; Mbaiwa 2005). Many of the tourist resorts in northern Finland, for example, have emerged as urban or semi-urban complexes with 25,000–40,000 bed places while the areas surrounding the same administrative region are characteristically rural or even uninhabited environments.

Thus, there is a need to re-think the conceptual nature of tourist destinations, especially if the images, representations, place promotion issues, and other processes socially constructing the spaces of tourism, are also seen as integral and crucial parts of the idea of destinations – as they should be (see Urry 1990; Shields 1991; Squire 1994; Edensor 1998). This call for re-thinking the destination is not a new issue in tourism studies: as argued by Mitchell (1984, 5), the geographical analysis of tourism has generally stressed the unique case rather than the general situation or theoreticization (see also Britton 1979; Ioannides 1996; Franklin and Crang 2001). Therefore, there is a need for conceptualizations integrating destinations with the larger regional structures and societal processes. As indicated by Viken (Chapter 2 in this volume), tourism development does not happen in a vacuum but can be seen as part of larger networks (see also Dredge 2006).

From a perspective of new regional geography (see Paasi 1991), one major implication is that destinations can be understood as subjects where universal and global processes and trends 'all come together' (see Keating 1998; Gordon and Goodall 2000, 292). The new regional geography (NRG), informed by structuration theory (see Giddens 1979), stresses the role of history, culture, social identities and power relations in the constitution of socio-spatial reality (Paasi 1986; Gilbert 1988). The NRG emphasizes the historical context and the interplay of spatiality with social reality, and spatial structures are seen as processes that are undergoing constant transformation. Thus, the focus lies not on the spatial structure itself in the first place but on the theoretical questions and empirical issues and their relations that construct and transform a region and connect with other regions and spatial scales (Paasi 1991).

In this respect the geographical concept of region offers a basis for defining, describing and analysing the tourist destinations as socially constructed spatial structures where global processes all come together and are manifested in a place in specific and concrete ways. The concept of region has gained various meanings during the history of academic geography (Paasi 1986), and some of them refer to the technical definitions of tourist destinations and tourism activities in bounded spaces (see Smith 1989, 163). However, as already stated, such definitions and approaches may leave out relational processes involved in constructing and historically forming destinations, their uses and their touristic nature (see Ryan 1991, 51; Burns and Holden 1995, 179; Saarinen 1998). Thus, tourist destinations as regions are not static units and their scale and characteristics are strongly related to the social production of space. From the viewpoint of the NRG, it is necessary to recognize the role of societal and institutional practices in constructing spatial structures in order to understand tourist destinations and their change and development (Saarinen 2004; see Gilbert 1988; Harvey 1993; Allen et al. 1998; Getz 1999). As a social construction, a tourism region becomes familiar to us through social processes and structures mediated by culture and communication (see Berger and Luckmann 1966; Paasi 1991).

Recently scholars have further stressed that regions are not entities separated from other geographical scales or socio-spatial processes (see Paasi 2009; 2012). Instead, they are seen as manifestations or outcomes of relations operating between different scales and regional units (Keating 1998). From this perspective, the tourist destination can be conceptualized as a historically produced structure which is lived, experienced and represented through different administrative, economic and cultural practices and through these practices the destination is constructed as a part of a larger regional (spatial) system and as part of the awareness of the tourists and public in general (Saarinen 2004). What this means is that the tourist destination can be understood as a socio-spatial construction characterized by the idea of region rather than physical or administrative bounds or elements.

Thus, a tourist destination is a reality that is produced and represented in a specific manner and it distinguishes the destination from its surrounding environment and other tourist destinations (Saarinen 1998; 2004). As Gilbert (1960, 158) has noted, 'regions, like individuals, have very different characters [which] … are constantly changing and developing'. Obviously there can be physical or territorial administrative boundaries between the destinations or the destinations and surrounding regions which may be important in some development and planning practices (see Smith 1989), but as already indicated for tourists or place promoters these boundaries – administrative and territorial limits – are not necessarily meaningful or even visible. Therefore, a more flexible and relational framework to analyse the nature and change of tourist destinations is needed which sees administrative (i.e. territorial) and relational geographical units as co-constitutive.

Transforming Tourist Destinations

Transformation process

The idea of tourist destination as a socio-spatial construction is not a dramatically new one, but the development process of destination has been a less conceptualized issue from this perspective (Saarinen 2004; see Davis 2001). From that perspective the development of tourism in certain spatial contexts can be understood as a part of larger social and ideological processes producing both the ideas and physical characters of destinations and the practices taking place in tourism development. These practices can be internal and external to the territory of destination and tourist activities. This complex set of changes, practices and relations leading to transforming socio-economic and cultural outcomes in development can be conceptualized through the idea of discourse (see Ateljevic and Doorne 2002). In general, discourses are socially and historically produced, coherent meaning systems and practices which both manifest and are power structures simultaneously (Wetherell and Potter 1992). While they are constructed in social practices, they also construct and reconstruct social reality as we perceive it but also the physical environment by virtue of the practices and policies attached to them. Thus, discourses are not 'just' talk and ideas as there are also material outcomes involved.

Thus, discourse as a term refers to both the process and its outcome. However, discourses are not static outcomes, they evolve with time (Crang 1998, 190). In this respect, they work well with the idea of tourist destination as constantly changing socio-spatial structure. In addition, there is not only one discursive statement or practice of a certain object, in this case tourist destination, but several. These multiple statements and actions create a discursive formation, constructing both knowledge and practices defining specific issues (Foucault 1980; see Hall 1992; 1997), such as the representation of a tourist destination or the limits of acceptable touristic changes at destinations (Saarinen 2004). While there may be several views and practices involved with destination development, one set of perspectives and actions is often hegemonic compared with others. Thus, there are always power issues involved in the representations and transformation processes of tourist destinations (Del Casino and Hanna 2000; see Hall 1994).

Discourses map and construct the idea of tourist destination that is meaningful for us, and discourses include the immaterial and material production of 'reality'.

This dual nature of discursive processes implies both the idea of tourist destination and the actions constructing the physical and symbolic landscape based on that idea(s), which can be analytically conceptualized through *a discourse of region* and *discourse of development* forming the identity of destination (Table 3.1) (Saarinen 2004; see Viken in Chapter 2 of this volume). The discourse of region refers to meanings, symbols and images related to tourist destinations, i.e., what we think the destination is about, while the discourse of development involves the practices and larger processes in the local-global nexus constructing destinations.

Table 3.1   The transformation of tourist destination as discursive practices

The Identity of Tourist Destination	
Discourse of region	*Discourse of development*
– representation, meanings and symbols	– institutional practices, planning and policy-making
– sources: advertisements, guide books, travel literature, television, internet etc.	– actors (internal/external): planning and development organizations and institutions and businesses

The discourse of region and development and related practices are highly inter-connected. The discourse of region includes the social and cultural meanings and materials creating the tourist destinations. These materials and related processes can include for example travel and regional literature, guide books and other place promotion materials and the media in general (see also Crang 1997; Thrift 1997, 190–194; Cloke and Perkins 1998, 187). These elements can be used as material in empirical research identifying the nature of tourist destinations (Hughes 1998). Thus, the discourse of region is based on the idea and conceptual nature of the region as a historical and social construction referring to a process in which the socio-spatial meanings and representations characterizing the destination are produced and reproduced (see Rojek 1997). This part of the transformation process creates and textualizes local geographies making the natural and cultural features of the destination known and popular – if successful. Thus, the idea of region – all that symbolizes and represents the tourist destination – is the main focus, and although the discourse refers to local scale it is an outcome of relations operating in a local-global nexus.

The discourse of development reflects the material and economic nature of tourism. This discourse involves the institutional processes shaping the destination by the practices of organizations and institutions set up for tourism development and planning. These practices and institutions operate in the destinations but they are not necessarily located in the destinations. In many cases the tourist destinations are shaped by different organizations and institutions working from the outside.

In empirical research the discourse of development can be characterized by policy and strategy documents and trends in the numbers of tourists, infrastructure and consumption, for example. However, these issues need to be understood as part of a larger societal and economic structure in which changes cannot be predicted causally from local manifestations at a certain destination. What this means, for example, is that a decline or rejuvenation (new cycle) in the development of a tourist destination is not necessarily a direct result of internal practices and outcomes in development actions, but can be based on larger regional or global economic situations and political relations causing changes in demand in the destination. Thus, the discourse of development links the destination inseparably to a larger regional and economic structure, and finally to the world economy and the circulation of capital. At the same time, however, it serves as an institutional tool and medium for the material development and construction of the identity of a region referring to a certain socio-spatial context.

The discourses of region and development together 'create' an identity for a tourist destination. The identity of destination is a *discursive formation* consisting of what the destination is and represents at the time and the historical and present practices involved in transforming it (Saarinen 2004). Thus, it includes the discourse of region with different texts and textual practices and the discourse of development with institutional processes and actions. This means that these discourses are co-constitutive, linking territorial and relational understandings of destination together as socially constructed spaces. As a social construct, the identity of a destination is a changing – temporally settled or unsettled – product of the process of transformation identifying the destination with both similar and distinctive elements relative to other tourist destinations.

Homogenization and differentiation

Tourist destinations are subject to the processes of discontinuities of modernity (see Giddens 1990) and currently tourist destinations are being transformed much more rapidly and on a more non-local basis than before (see Britton 1991; Lew 1999; Davis 2001). As a result of the transformation process, tourist destinations are often seen as developing towards spatial homogenization (Saarinen 2004; see Relph 1976) in which destinations evolve into places with similar physical and symbolic characteristics. This homogenization process demonstrates the idea of *time-space compression* (Harvey 1985; 1989) in tourism development in which space and spatial experience shrink as a result of increasing movement of capital, goods and information. As time and space are compressed and transformed by this circulation and over accumulation of capital, the nature of the tourist destination may also turn towards homogenization and mass-scale industry in order to serve more effectively the accumulation of capital.

Indeed, tourist destinations are spaces designed on the premises of attracting non-local capital and tourists (Getz 1999). From that perspective the spatial homogenization is a logical process and outcome in tourism development. The homogenization process is based on stereotyping production of representations, services and other facilities often involving unequal power relations between different stakeholders such as the tourism industry, other livelihoods and local communities (see Cheong and Miller 2000, 372; Pritchard and Morgan 2000, 116). Finally, the process of spatial homogenization may lead to spatial differentiation. This may sound contradictory but it originates from the very same transformation process: while destinations are growing and transforming towards similarity, i.e., homogenization, they are also undergoing differentiation and being separated from their surrounding regional structure (and their 'original' i.e. previous characteristics (Saarinen 2004; see also MacCannell 1976, 44–48; Rojek 1993).

The homogenization and differentiation can lead to the evolution of enclavic tourism spaces which have limited connections to surrounding regional structures and processes. As such they are products of regulated spaces and bordering practices by private organizations and/or public sector actors. Thus, while there are increasing 'rights of mobility' (Turner 2007) between places and national and international scales in the contemporary world, there are also processes challenging the fluidity of movement in tourism and tourist destinations. As noted by Britton (1991, 452), tourism and tourists are not 'free in any absolute sense' but subject 'to rules of permissible forms and sanctioned behaviour'. This notion obviously includes also local populations in host-guest relations and encounters (see Gibson 2011) creating spaces that are only for visitors and their consumption needs (see also Getz 1999; Hollinshead 1999). Indeed, the homogenization and differentiation of tourist destinations can involve elements creating complex bordering practices where some people can enter the destination spaces or certain parts of them while the movement of others is limited (see Carlisle and Jones 2012), e.g. in some places, especially in developing countries, only tourists are allowed to use certain beach areas, or lands formerly used for grazing local people's cattle are turned into private game reserves for paying wildlife tourists. Britton's (1991) enclave model, for example, states that there are regulatory controls and hierarchical relations of power and flows of capital, knowledge, good and services in destination development. While Britton's case is placed in the context of developing countries the same applies, although in a less dramatic scale, almost in any given peripheral settings in Western world (see Ringer 1998; Saarinen 2003; 2004).

Evolutionary approaches to transformation process

Recently, new approaches related to evolutionary perspectives to socio-economic development have received increased attention in destination and regional development studies, especially in human and economic geography (see Williams and Baláž 2000; 2002; Martin and Sunlay 2006), and in tourism geographies (see Saarinen and Kask 2008; Ma and Hassink 2013). In addition to regional policy implications, networks and their evolution and relations, entrepreneurship, innovations or firm dynamics and survival, and governance (see Floysand and

Jakobsen 2010; Hjalager 2010), for example, the aspects of path-dependency and path-creation in regional and tourism development have gained interest in research (Baláž and Williams 2005). The dimensions of path-dependency and path-creation in destination development help to understand and interpret the local responses to internal and external processes, structures and changes. Some of the current processes and challenges that are highlighting the discussions on path-dependency and path-creation in destination development are global financial crisis, internationalization of customer base, global climate change adaptation and mitigation needs, and sustainability (Scott 2011; Bramwell and Lane 2012; Kaján and Saarinen 2013).

In general, path-dependence refers to a development situation in which the actors and their possibilities are relatively firmly limited by existing structures and resources. These resources and structures support some development strategies (i.e., paths) and actions much better than others (Williams and Baláž 2002) often reflecting the hegemonic discourses of the transformation process of tourist destinations. In contrast, path-creation stresses the active role of local actors who can, within certain realistic limits, influence and modify the course of local and regional development and, thus, innovate and avoid a lock-in situation in development (Nielsen et al. 1995; Stark 1996; Saarinen and Kask 2008). These new pathways can potentially represent conflicting views to the hegemonic discourse of what the destination is and represents. Thus, while the present identity of destination is characterized mainly by hegemonic views it also contains features from the past and signs of future transformations of the destination which can be competing with each other and with the hegemonic views (Saarinen 2004). This represents a similar perspective to the idea of hegemonic culture discussed by Raymond Williams (1988), which involves not only the current dominant understanding what 'we are' but also traces and values from the past, such as displaced traditions and ways of using environment, and emerging pathways and 'weak signals'.

New evolutionary perspectives on tourist destination development are focusing on the analysis of the changing characteristics and underlying reasons of destination development in a changing operational environment (Ma and Hassink 2013). This changing environment can be ecological, socio-cultural or economic but also political, involving neoliberalization and the emergence of new governance with increasing privatization and cuts in the public sector support for regional development and policy activities (see Rhodes 1996). In this respect a key issue relates to the question why some places become locked into certain negative development paths, narrowing their future growth potential and options, whilst other places avoid the same trap and manage to operate in a changing operational environment (Martin and Sunlay 2006, 395). In this sense, analyses of the transformation process and identification of alternative paths could be used as a tool for considerations of future tourism development in certain tourist destinations.

Discussion and Conclusions

The idea of tourist destination as a social construction is not a new one. The changing nature of a tourist destination, however, has been a less analysed and conceptualized issue from this perspective. As social constructions tourist destinations are changing products of social forces and relations. On the other hand, these social relations and realities are also influenced and organized by geographical structures in certain socio-spatial contexts which make tourist destinations as dynamic and historical conditioned regional units. As such the destinations can be seen as having a specific identity characterized by hegemonic discourses, which produce a notion of what the destination is and represents at the time. Thus, tourist destinations are not distant locations 'out there' but ours and others' constructions (Saarinen 2004; see Allen et al. 1998, 2). From this perspective destinations are seen as constantly changing products of a certain combination of social, political and economic relationships that are specific in space and time.

While deepening globalization moulds destinations more drastically than ever before, the processes that we link to the idea of globalization (see Robertson 1992; Harvey 1993) are not solely driven from 'out there', i.e., one-way from global to local, as local actors can also contribute to the processes of globalization and especially their local outcomes (see Giddens 1998; Teo 2002). Thus, even though globalization and global climate change, for example, are often seen as external and strictly beyond local scale, local perspectives and actions can have a key role to play: also externally oriented processes are mediated by and their impacts and implementation are influenced by local actors (see Massey and Allen 1984; Massey 1991; Teo and Li 2003). Obviously this capacity to relate actively with global forces does not make global-local relations equal in tourism or more generally in development. As noted by Lewandowski (2003, 126), 'those who in fact experience globalization on the ground come to discover rather quickly ... new forms of spatial and temporal control'. This raises issues of resistance and resilience (see Falk 1997) and also highlights the role of local knowledge and human agency in path creation in transforming destinations.

The idea of tourism development as a discursive process aims to provide a framework for analysing the nature and change of tourist destination. The discourses of region and development are conceptual views and the distinction between the concepts is obviously an analytical one. Thus, while the discourses and their ideological and practical dimensions are all interlocked, analytically it is possible to separate the institutional practices from the socio-spatial representations and meanings of destination. The discourse of region signifies and spatializes the practices of development, while the discourse of development produces media and infrastructures for place-making, for example.

While tourist destinations are distinguishable from their surrounding regional structures and from other destinations based on the idea and identity of region, they are not static or internally homogeneous. There can be competing discourses and conflicting elements of identities and meanings, and not only one after another, but also concurrently (Saarinen 1998; 2004). Developing tourism may often produce destination identities representing the values and needs of the (non-local or local) tourism industry rather than other local interests or identities (Urry 1990, 64). From that perspective the analysis of the transformation of tourist destinations should consider not only the hegemonic but also the challenging and conflicting discourses (see Viken in Chapter 2 of this volume). This is crucial especially in peripheral areas where tourism often seems to represent a development 'that satisfies the commercial imperatives of an international business, yet rarely addresses local development needs' (Ringer 1998, 9). Therefore, in respect to sustainable tourism development the non-hegemonic views and groups involved with transforming tourist destinations should be identified, allowed to participate and empowered in development. In addition, a wider spectrum of views to symbolic and material practices steering the development should be considered along the hegemonic perspectives: that would probably be more sustainable but it would also empower a path-creation in discussions and strategies on destination development.

References

Agarwal, S. (2002), 'Restructuring Seaside Tourism: The Resort Lifecycle', *Annals of Tourism Research* 29:1, 25–55.

Aitchison, C., MacLeod, N.E. and Shaw, S.J. (2000), *Leisure and Tourism Landscapes: Social and Cultural Landscapes* (London and New York: Routledge).

Allen, J. et al. (1998), *Rethinking the Region* (Routledge: London).

Ashworth, G.J. and Dietvorst, A.G.J. (eds) (1995), *Tourism and Spatial Transformations* (Oxon: CAB International).

Ateljevic, I. and Doorne, S. (2002), 'Representing New Zealand: Tourism Imagery and Ideology', *Annals of Tourism Research* 29:3, 648–667.

Baláž, V. and Williams, A. (2005), 'International Tourism as Bricolage: An Analysis of Central Europe on the Brink of European Union Membership', *International Journal of Tourism Research* 7:2, 79–93.

Bird, J. et al. (eds) (1993), *Mapping the Futures: Local Cultures, Global Change* (London: Routledge).

Bramwell, B. and Lane, B. (2012), 'Towards Innovation in Sustainable Tourism Research', *Journal of Sustainable Tourism* 20, 1–7.

Burger, P. and Luckmann, T. (1966), *The Social Construction of Reality: A Treatise in the Sociology of Knowledge* (London: Allen Lane).

Britton, R. (1979), 'Some Notes on the Geography of Tourism', *Canadian Geographer* 23:3, 276–82.

Britton, S.G. (1991), 'Tourism, Capital, and Place: Towards a Critical Geography of Tourism', *Environment and Planning D: Society and Space* 9: 451–78.

Burns, P.M. and Holden, A. (1995), *Tourism. A new Perspective* (London: Prentice Hall).

Butler, R. (1980), 'The Concept of a Tourist Area Cycle of Evolution: Implications for Management of Resources', *Canadian Geographer* 24:1, 5–12.

Carlisle, S, and Jones, E. (2012), 'The Beach Enclaves: A Landscape of Power', *Tourism Management Perspectives* 1:1, 9–16.

Cheong, S.-M. and Miller, M.L. (2000), 'Power and Tourism: A Foucauldian Observation', *Annals of Tourism Research* 27:2: 371–90.

Christaller, W. (1963), 'Some Considerations of Tourism Location in Europe: The Peripheral Regions-under Development Countries-Recreation Areas', *Regional Science Association; papers XII, Lund Congress*, 95–105.

Cloke, P. and Perkins, H. (1998), '"Cracking the Canyon with the Awesome Foursome": Representation of Adventure Tourism in New Zealand', *Environment and Planning D: Society and Space* 15:2, 185–218.

Cooper, C. (1989), 'Tourist Product Cycle', in Witt and Moutinho (eds).

Crang, M. (1997), 'Picturing Practice: Research through the Tourist Gaze', *Progress in Human Geography* 21:3, 359–373.

—— (1998), *Cultural Geography* (London: Routledge).

Crang, M. and Thrift, N. (eds) (2000), *Thinking Space* (London and New York: Routledge).

Davis, J.B. (2001), 'Commentary: Tourism Research and Social Theory – Expanding the Focus', *Tourism Geographies* 3:2, 125–134.

Del Casino, V.J.Jr. and Hanna, S.P. (2000), 'Representation and Identity in Tourism Map Spaces', *Progress in Human Geography* 24:1, 23–46.

Dredge, D. (2006), 'Policy Networks and the Local Organisation of Tourism', *Tourism Management* 27:2, 269–280.

Edensor, T. (1998), *Tourists at the Taj: Performance and Meaning at a Symbolic Site* (London and New York: Routledge).

Eurostat (2013).NUTS – Nomenclature of territorial units for statistics <http://epp.eurostat.ec.europa.eu/portal/page/portal/nuts_nomenclature/introduction> European Commission.

Falk, R. (1997), 'Resisting "Globalisation-from-Above" through "Globalisation-from-Below"', *New Political Economy* 2:1, 17–24.

Featherstone, M. and Lash, S. (1995), 'Globalisation, Modernisation and the Spatialisation of Social Theory', in Featherstone and Robertson (eds).

Featherstone, M. and Robertson, R. (eds) (1995), *Global Modernity* (London: Sage).

Fløysand, A. and Jacobsen, S.-E. (2010), 'The Complexity of Innovation: A Relational Turn', *Progress in Human Geography* 35:3, 328–344.

Foucault, M. (1980), *Power/Knowledge: Selected Interviews and Other Writings 1972–1977 by Michel Foucault* (ed. by C. Gordon) (New York: Harvester Wheatsheaf).

Framke, W. (2002), 'The Destination as a Concept: A Discussion of the Business-Related Perspective versus the Socio-Cultural Approach in Tourism Theory', *Scandinavian Journal of Hospitality and Tourism* 2:2, 91–108.

Franklin, A. and Crang, M. (2001), 'The Trouble with Tourism and Travel Theory', *Tourist Studies* 1:1, 5–22.

Gale, T. and Botterill, D. (2005), 'A Realist Agenda for Tourist Studies, or Why Destination Areas Really Rise and Fall in Popularity', *Tourist Studies* 5:2, 151–174.

Getz, D. (1999), 'Resort-Centred Tours and Development of the Rural Hinterland: The Case of Cairns and the Atherton Tablelands', *The Journal of Tourism Studies* 10:2, 23–34.

Giddens, A. (1979), *Central Problems in Social Theory: Action, Structure and Contradiction in Social Analyses* (London: Macmillan).

—— (1990), *The Consequences of Modernity* (Standford: Standford University Press).

—— (1998), *The Third Way: The Renewal of Social Democracy* (Cambridge: Polity Press).

Gibson, C. (2010), 'Geographies of Tourism: (Un)ethical Encounters', *Progress in Human Geography* 34, 521–527.

Gilbert, A. (1988), 'The New Regional Geography in English and French-Speaking Countries', *Progress in Human Geography* 12:2, 208–228.

Gilbert, E.W. (1939), 'The Growth of Island and Seaside Health Resorts in England', *Scottish Geographical Magazine* 55:1, 16–35.

—— (1949), 'The Growth of Brighton', *The Geographical Journal* 114:1/3, 30–52.

—— (1960), 'The Idea of the Region', *Geography* 45, 157–175.

Gordon, I. and Goodall, B. (2000), 'Localities and Tourism', *Tourism Geographies* 2:3, 290–311.

Gregory, D. and Urry, J. (eds) (1985), *Social Relations and Spatial Structure* (London: Macmillan).

Hall, C.M. (1994), *Tourism and Politics: Policy, Power and Place* (Chichester: John Wiley & Sons).

—— (1997), 'Geography, Marketing and the Selling of Places', *Journal of Travel and Tourism Marketing* 6:3–4, 61–84.

Hall, C.M. et al. (2009), *Nordic Tourism: Issues and Cases* (Bristol: Channel View Publications).

Hall, C.M. and Saarinen, J. (2010), 'Tourism and Change in the Polar Regions: Introduction – Definitions, Locations, Places and Dimensions', in Hall and Saarinen (eds).

Hall, C.M. and Johnston, M.E. (eds) (1995), *Polar Tourism: Tourism in the Arctic and Antarctic Regions* (Chichester: John Wiley & Sons).

Hall, C.M. and Saarinen, J. (eds) (2010), *Tourism and Change in Polar Regions: Climate, Environments and Experiences* (London: Routledge).

Hall, S. (1992), 'The West and the Rest', in Hall and Gieber (eds).

—— (1997), 'The work of representation', in Hall (ed.).

—— (ed.) (1992), *Representation: Cultural Representations and Signifying Practices* (London: The Open University and Sage).

Hall, S. and Gieber, B. (eds) (1992), *Formations of Modernity* (London: Polity Press and The Open University).

Harvey, D. (1985), 'The Geopolitics of Capitalism', in Gregory and Urry (eds).

—— (1989), *The Condition of Postmodernity* (Oxford: Blackwell Publishers).

—— (1993), 'From Space to Place and Back Again: Reflections on the Condition of Postmodernity', in Bird et al. (eds).

Hausner, J. et al. (eds) (1995), *Strategic Choice and Path Dependency in Post Socialism: Institutional Dynamics in the Transformation Process* (Aldershot: Edward Elgar).

Haywood, K.M. (1986), 'Can the Tourist-Area Life Cycle be Made Operational?' *Tourism Management* 7:3, 154–167.

Hjalager, A.-M. (2010), 'Regional Innovation Systems: The Case of Angling Tourism', *Tourism Geographies* 12:2, 192–216.

Hollinshead, K. (1999), 'Surveillance and the Worlds of Tourism: Foucault and the Eye of Power', *Tourism Management* 20:1, 7–24.

Hughes, G. (1998), 'Tourism and the Semiological Realisation of Space', in Ringer (ed.).

Ioannides, D. (1996), 'Tourism and Economic Geography Nexus: A Response to Anne-Marie d'Hauteserre', *Professional Geographer* 47:1, 219–221.

Johnston, C.S. (2001), 'Shoring the Foundations of the Destination Life Cycle Model, Part 1: Ontological and Epistemological Considerations', *Tourism Geographies* 4:1, 2–28.

Kajan, E. and Saarinen J. (2013), 'Tourism, Climate Change and Adaptation: A Review', *Current Issues in Tourism* 16:2, 167–195.

Keating, M. (1998), *The New Regionalism in Western Europe* (Cheltenham: Edward Elgar).

Leiper, N. (1990), 'Tourist Attraction System', *Annals of Tourism Research* 17:3, 367–384.

Lew, A. (1987), 'A Framework of Tourism Attraction Research', *Annals of Tourism Research* 14:4, 553–575.

—— (1999), 'Editorial: A Place Called Tourism Geographies', *Tourism Geographies* 1:1, 1–2.

Lewandowski, J.D. (2003), 'Disembedded Democracy? Globalization and the "Third Way"', *European Journal of Social Theory* 6: 115–131.

Ma, M. and Hassink, R. (2013O), 'An Evolutionary Perspective on Tourism Area Development', *Annals of Tourism Research* 41, 89–109.

MacCannell, D. (1976), *The Tourist. A New Theory of the Leisure Class* (New York: Schoken Books).

Mackay, H. (ed.) (1997), *Consumption and Everyday Life* (London: Sage).

Martin, R.L. and Sunlay, P.J. (2006), 'Path Dependency and Regional Economic Evolution', *Journal of Economic Geography* 6, 395–438.

Massey, D. (1991), 'The Political Place of Locality Studies', *Environment and Planning A* 23: 267–281.

Massey, D. and Allen, J. (eds) (1984), *Geography Matters! A Reader* (New York: Cambridge University Press).

Mbaiwa, J.E. (2005), 'Enclave Tourism and its Socio-economic Impacts in the Okavango Delta, Botswana', *Tourism Management* 26:2, 157–172.

Meethan, K. (1996), 'Place, Image and Power: Brighton as a Resort', in Selwyn (ed.).

Meethan, K. (2001), *Tourism in Global Society: Place, Culture, Consumption* (Basingstoke: Palgrave).

Miossec, J. (1967), 'Un modele de l'espace touristique', *L'Espace Geographique* 6:1, 4–8.

Mitchell, L. (1984), 'Tourism Research in the United States: A Geographic Perspective', *GeoJournal* 9:1, 5–15.

Nielsen, K. et al. (1995), 'Institutional Change in Post-Socialism', in Hausner et al. (eds).

Paasi, A. (1986), 'The Institutionalization of Regions: A Theoretical Framework for Understanding the Emergence of Regions and Constitution of Regional Identity', *Fennia* 164:1, 105–146.

—— (1991), 'Deconstructing Regions: Notes on the Scales of Spatial Life', *Environment and Planning A* 23, 239–256.

—— (2009), 'The Resurgence of the 'Region' and 'Regional Identity': Theoretical Perspectives and Empirical Observations on Regional Dynamics in Europe', *Review of International Studies* 35, 121–146.

—— (2012). 'Regional Planning and the Mobilization of 'Regional Identity': From Bounded Spaces to Relational Complexity', *Regional Studies*, DOI: 10.1080/00343404.2012.661410. (iFirst article).

Pritchard, A. and Morgan, N.J. (2000), 'Constructing Tourism Landscapes – Gender, Sexuality and Space', *Tourism Geographies* 2:2, 115–139.

Relph, E. (1976), *Place and Placelessness* (London: Pion).

Rhodes, R.A.W. (1996), 'The New Governance: Governing without Government', *Political Studies* XLIV, 652–667.

Ringer, G. (1998), 'Introduction', in Ringer (ed.).

Ringer, G. (ed.) (1998), *Destinations: Cultural Landscapes of Tourism* (London and New York: Routledge).

Robertson, R. (1992), *Globalisation: Social Theory and Global Culture* (London: Sage).

Rojek, C. (1993), *Ways of Escape. Modern Transformations in Leisure and Travel* (London: MacMillan).

—— (1997), 'Indexing, Dragging and the Social Construction of Tourist Sights', in Rojek and Urry (eds).

Rojek, C. and Urry, J. (eds) (1997), *Touring Cultures: Transformations of Travel and Theory* (London and New York: Routledge).

Ryan, C. (1991), *Recreational Tourism: A Social Science Perspective* (London: Routledge).

Saarinen, J. (1998), 'The Social Construction of Tourism Destination: The Transformation Process of Saariselkä Resort in Finnish Lapland', in Ringer (ed.).

—— (2003), 'The Regional Economics of Tourism in Northern Finland: The Socio-Economic Implications of Recent Tourism Development and Future Possibilities for Regional Development', *Scandinavian Journal of Hospitality and Tourism* 3:2, 91–113.

—— (2004), '"Destinations in Change": The Transformation Process of Tourist Destinations', *Tourist Studies* 4:2, 161–179.

Saarinen, J. and Kask, T. (2008), 'Transforming Tourism Spaces in Changing Socio-Political Contexts: The Case of Pärnu, Estonia, as a Tourist Destination', *Tourism Geographies* 10:4, 452–473.

Scott, D. (2011), 'Why Sustainable Tourism Must Address Climate Change', *Journal of Sustainable Tourism* 19, 17–34.

Selwyn, T. (ed.) (1996), *The Tourist Image: Myths and Myth Making in Tourism* (Chichester: John Wiley & Sons).

Shields, R. (1991), *Places on the Margin: Alternative Geographies of Modernity* (London: Routledge).

Smith, S.L.J. (1989), *Tourism Analyses: A Handbook* (London: Longman).

Squire, S.J. (1994), 'Accounting for Cultural Meanings: The Interface between Geography and Tourism Studies Re-examined', *Progress in Human Geography* 18:1, 1–16.

Stark, D. (1996), 'Recombinant Property in East European Capitalism', *American Journal of Sociology* 101: 4, 933–1027.

Teo, P. (2002), 'Striking a Balance for Sustainable Tourism: Implications of the Discourse on Globalisation', *Journal of Sustainable Tourism* 10:6, 459–74.

Teo, P. and Li, H.M. (2003), 'Global and Local Interactions in Tourism', *Annals of Tourism Research* 30, 287–306.

Terkenli, T. (2002), 'Landscapes of Tourism: Towards a Global Cultural Economy of Space?', *Tourism Geographies* 4:3, 227–254.

Thrift, N. (1997) '"Us" and "Them": Re-imagining Places, Re-imagining Cultures', in Mackay (ed.).

Turner, B.S. (2007), 'The Enclave Society: Towards a Sociology of Immobility', *European Journal of Social Theory* 10, 287–303.

Urry, J. (1990), *The Tourist Gaze: Leisure and Travel in Contemporary Societies* (London: Sage).

—— (1995), *Consuming Places* (London: Routledge).

Viken, A. (1995), 'Tourism Experiences in the Arctic – the Svalbard Case', in Hall and Johnston (eds).

Wetherell, M. and Potter, J. (1992), *Mapping the Language of Racism: Discourse and Legitimation of Exploitation* (Cornwall: Harvester Wheatsheaf).

Williams, A.M. and Baláž, V. (2000), *Tourism in Transition* (London and New York: I.B.Tauris).

Williams, A.M. and Baláž, V. (2002), 'The Czech and Slovak Republics: Conceptual Issues in the Economic Analysis of Tourism in Transition', *Tourism Management* 23:1, 37–45.

Williams, R. (1988), *Keywords: A Vocabulary of Culture and Society* (Fontana Press, London).

Witt, S. and Moutinho, L. (eds) (1989), *Tourism Marketing and Management Handbook* (London: Prentice Hall).

Chapter 4

Destination Development Performances: Or How we Learn to Love Tourism

Simone Abram

This chapter takes a journey through tourism studies to explore what the destination might be. Should destination management research only analyse 'destination management' organizations? Can tourism research consider destination development without the contribution of research on development? What is there to learn about destinations from planning research? How can theoretical debates in the social sciences inform approaches to destination research? By considering recent debates over 'performativity', the chapter shifts the focus first to the *framing* of destination research, then to asking *how we learn* to love tourism, before considering how the destination is *produced* in practice as one of many kinds of development.

Finding the Destination

Walking through a heritage visitor centre recently to access a row of houses on the other side, I was reminded of the peculiarity of the tourism performance. As a person walking through, rather than visiting, I was clearly out of kilter with the other people at the site. Walking quickly, to avoid being late visiting relatives at the house, walking past rather than around the various attractions at the site, and with my wallet securely in my pocket, I could easily have been spotted as someone who had ignored the 'patrons only' sign at the car park. Being out of place offers a critical position to question what everyone else was doing, why visitors tend to stroll rather than race, what attracts people to visit these places, and what they do when they are there. Similar questions arise at all tourism destinations, and tourism studies of various disciplines have explored many of these questions over the years. What 'motivation studies' and 'attitude studies' offer is a series of hierarchically organized factors representing the consciously articulated reasons offered by visitors for their choice of destination and activity. What they tell us less about, however, is how these motivations appeared, why they are articulated in particular ways, or how people's choices are structured prior to them being made. In this chapter, I suggest that debates about performativity and attention to the material affordances of tourism infrastructure can help us develop a line of argument in tourism research about the development of desire, or how we learn to love tourism.

Since Urry's notable Foucauldian analysis of tourism in 1990, tourism researchers have been aware that tourism exists in a scheme of social expectation, norms and tacit knowledge. Urry pointed out that tourism was not merely a set of activities, a procedure involving journeying, or a way of selling experience. On the contrary, we could understand tourism as a peculiarly modern way of experiencing and interpreting the world around us, and of acting within it. As Mitchell (1988) has indicated, tourism was a crucial factor in the colonizing process, and the way European tourists looked at Egypt was shaped by a set of modernizing practices honed through the great exhibitions of London and Paris. Egypt was presented to visitors to the exhibition as a spectacle to be consumed by the senses, predominantly vision, so that tourists who then journeyed to Egypt went in expectation of experiencing something similar. Experiencing the world as an exhibit presented for the pleasure of the travelling consumer became the archetypal modernist and colonial approach to the world, suggesting, as Franklin (2004) has argued, that tourism is a form of ordering the world, with its own range of ordering effects. As he recognizes, a person does not become a tourist spontaneously or independently, but first encounters the possibility of tourism, considers the journey and the destination as attractive possibilities, and subsequently relies on a complex array of interlocking governmental, technological, social and organizational systems, in order to be able to put the imagined journey into motion. Desire for a holiday or other travel cannot be explained as a simple product of marketing techniques, but is prefigured and conditioned by larger societal frames that are emotional and cognitive as much as technological and economic. Alongside theories of human behaviour, understanding tourism requires us to have theories of global change, and at the same time offers us a means to explore and theorize the changes represented by the terms modernism, colonialism, post-colonialism, neo-liberalization and contemporary globalization.

At the same time, we cannot study tourism in practice if our sights remain on the broad global stage of the world and macro-historic scales of change. Such debates are meaningful only if they are based on what we know about tourism practices, which, by definition, always happen in particular places and times (as Mitchell's historical analysis demonstrates). Picard's detailed ethnographic evidence of tourism practices, for example, leads him to a psychoanalytical approach in which he considers the cultivation of tourism as a form of enchantment. The literal and metaphorical cultivation of gardens for tourists represents for him a broader cultivation of sites and places of tourism.

> The long-standing and widely common practice of contemplative gardening – cultivating miniaturized models of the world, invoking and thus bringing into being ideas of a wider cosmos, working with mimetic or metonymic matter related to such a cosmos – has flowed into the realms of international policy making and tourism development in particular. (Picard 2011, 5)

As productive agriculture has failed in successive regions, another kind of gardening has replaced it: what Picard calls the contemplative 'human gardens' of new world society, in which the middle classes recreate the symbolic order of their worlds (ibid.). The transformation of the island of La Reunion from a site of colonial production to a beautiful tropical island has implied a transformation of aesthetics and of ethics for islanders as well as for visitors, and Picard argues that such transformations adopt the practices of mimesis, invocation, transfiguration and metonym to enhance the magical qualities that tourists perceive on the island. Magical practices maintain the separation between self and other, so that the island and its population can remain an exotic attraction for the benefit of tourists, reinforcing their preconceptions about the archetypal tropical island paradise and its people.

As theoreticians, we cannot understand the trope of the tropical island paradise without the broader frame of European historical philosophy, yet we cannot expect that tourists have a particularly high level of knowledge of that philosophy. On the contrary, the majority of both tourists and tour operators adopt tropes and symbols with minimal reflection. This is, after all, how symbols work, as shorthand identifiers of sometimes highly complex ideas. It is enough to construct an image of the sun setting over the far ocean horizon, with beach and palm trees in the foreground, with or without a good looking European couple gazing fondly on, to evoke the history of imaginations of paradise from bible teachings to Swiftian satire, from chick lit to adventure films.[1] Island paradise is only one of the many symbols adopted in the cultivation of tourism sites, be they resorts, attractions, destinations or centres. Tourism operators and marketers have developed a range of tropes that are soon taken for granted by operators and tourists alike, immune to sociological deconstruction. Wilderness landscapes, remote outposts, stone-age peoples, authentic communities, exciting urban quarters, tranquil retreats, and so forth, can be seen as identifiers of an order to which tourism predominantly adheres. The point to make here is that such symbols are not merely there, but are invented, presented, received, circulated, interpreted and adopted. Deconstructions of the rhetorics and semiotics of tourism signs and discourses are valuable tools to reveal the meanings that are otherwise hidden, but they are only half the story. We need also to understand how people *do* tourism, how symbols are put into motion, what actions they prompt, and what effects they have.

One of the ways that action has been restored to tourism studies has been through a focus on the body (Veijola and Jokinen 1994). Paying attention to the bodily functions of tourism, the body of the tourist and the embodiment of tourism knowledges has been a crucial way to return our focus to the physical rather than the purely discursive and to remind us that material and corporeal forms help to sustain and reproduce broader discourses. Equally, the influence of socio-technical approaches to tourism inspired by the interdisciplinary science studies methods known amongst other things as material semiotics (Law 2008) has been an important means to acknowledge the interplay of symbol, practice

---

1   A google websearch of 'island paradise' offers nearly 56 million results.

and performance in tourism. A flourishing literature on the material semiotics of sport fishing and hunting has led the way in this regard (see Abram and Lien 2011 special issue articles).

Learning to be a Tourist

Where tourism research has been less attentive has been in questioning how this form of ordering is facilitated materially and how it is absorbed. How does the governance of development generate certain kinds of destination? How do people know what an ideal holiday should be? How do we internalize the desire to participate in certain activities? In short, how do we learn to be a tourist?

Formal learning is not a major factor in our learning to be tourists, although one might identify commonly organized school trips to sites of interest, such as museums, learning centres or sites of historical or heritage interest[2] as moments when children begin to learn to be tourists. These trips teach us from an early age that there are places where we can learn outside formal educational premises, and that we can enjoy a 'day out' from normal service, and that this should be seen as a pleasure or a treat.

Much more of our learning about tourism is informal. We pick up information passively from different sources of media and marketing. Holidays offered as prizes in competitions, as work incentives, or as prizes from television shows and media competitions help to reinforce the idea of tourism products as desirable, for example. Perhaps most importantly, we learn from doing; we are introduced to ways of being a tourist by family and friends. Being taken on holiday as children by our parents, for example, can be a source of an adult's nostalgic desire to recreate childhood pleasures, or as an antithesis for the kind of tourist we want to be as adults. Similarly, talking about holidays at the hairdressers is a way of doing non-committal sociality, while reinforcing the position of a holiday as a normal focus of desire. This kind of learning is characterized in one field of educational theory as learning by situated doing. According to Jean Lave and Etienne Wenger, much of our learning happens through a process of deepening participation in established practices (1991). Whether that is as a child being socialized into a family or wider community context, or as an adult becoming part of what Wenger describes as a 'community of practice' (1998), we seldom learn in isolation. Lave and Wenger's approach to collectively situated learning provides a counter to theories of individual learning and serves to remind us that the ethnographer's way – moving from a peripheral to a gradually more central position in a community as we learn how to behave appropriately and how to be – is parallel to much social learning

---

2   School trips inevitably go in and out of fashion depending on availability of funding and the flexibility of curricula. I have in mind northern European school parties, although, of course, school trips are not limited to European schools.

(see Evans 2006). Ethnographers call on familiar means of negotiating the world through the organization and assimilation of knowledge:

> Being alive as human beings means that we are constantly engaged in the pursuit of enterprises of all kinds, from ensuring our physical survival to seeking the most lofty pleasures. As we define these enterprises and engage in their pursuit together, we interact with each other and with the world and we tune our relations with each other and with the world accordingly. In other words we learn. (Wenger 1998, 45)

In doing so, the resulting practices both constitute the communities in which we perform them, and the social relations which support the practice. In the tourism context, it is useful to ask which communities of practice can be identified, since there are many potential environments in which practices are developed, and these interweave in the context of mobility that tourism represents. While on the one hand, we learn to generate tourism desires through peripheral participation as members of a family or social group or work organization, our desires for tourism are both met and manipulated by tourism developers offering and promoting particular forms of tourism, themselves learning and innovating through a different set of practice communities. The focus on practice reveals the processes of learning that are undertaken, but they also shift the research focus from philosophical considerations of travel and tourism to the empirical experiences of those involved.

This renewed focus in the social sciences on performativity refocuses attention from what critics have seen as an obsession with representations and discourses (e.g. Barad 2003). While we can deconstruct the language and categorizations used in describing what we do, Barad argues against a fixation on words and things in favour of practices and relations. Rather than finding the gaps between what is said and what is done, the performative idiom analyses how models and practices work together and the specific materials they create. It builds also on the philosophy of linguistics that recognizes that words are not simply there but are spoken, written and read, and gain their power from their utterance. Words thus come alive through the relations they create between speaker and listener, writer and reader, and between thinking and doing. Performativity develops the remark by Austin that an utterance is the performance of an action, and his identification of speech acts as creative actions in their own right (Austin 2009/1962).

Research on performance in tourism, however, emerges largely from a different tradition of performance research, as Harwood and El Manstrly observe in their observation of articles in Annals of Tourism Research (2012). Tourism studies tend to draw on Goffmanesque theatrical metaphor, and the anthropological tradition of understanding ritual through theories of drama and staging (especially from Turner 174; see McCannell 1976; 1992; Bærenholdt et al. 2004). This has been a rich source of inspiration for understanding the roles and performances that tourism demands, and helps us to understand that things are not always as they seem to tourists, for example. It is also the source of a great deal of speculation, discussion and, indeed, anxiety around the concept of 'authenticity'. In contrast, the notion of performativity examines how the categories we think and speak with are generated through the total performance of those categories, in the utterance of speech and the doing of actions that conform to those categories.[3] We not only discuss tourism but act as though it exists. Once the idea of a destination as an object of management appeared, destination managers began to try to operationalize the destination, and their continued efforts maintain the existence of that destination as an object of speech, action, speculation and policy activity.

If this approach appears to be increasingly abstract, it is usefully brought back down to earth through a complementary approach that builds on the same theoretical premise. The material world is understood through communication – both linguistic and physical – but also shapes that communication. We cannot walk through walls, however much we may try to convince ourselves that it is possible. On the contrary, the material conditions of our existence afford us certain possibilities and not others. Hence, a focus on the 'affordances' of the material world complements the linguistic approach to performativity, and demands that we take seriously the non-human agency of the material world, in our attempts to explain why things are the way they are. This combination of performance, performativity and material ontology offers the opportunity to look with a more discerning eye at what is actually happening in tourism destinations and destination development and to re-evaluate the variety of work that takes place to keep tourism moving, whether this is the creation of desire or the technologies that enable people to move from place to place.

How is Tourism Done?

With this in mind, the questions we can ask revolve around how tourism is done. Juxtaposed with Picard's observations on the cultivation of magical landscapes that seduce tourists, we might ask how tourists get to the point of indulging their fantasies of exotic island paradise. How, for example, is tourism infrastructure brought into being? Callon (1991) has pointed out the 'machinic' side of travel, the implication of motors and machines in getting us around, and the complex social and organizational relations that enable us to make, use and master these machines. A package holiday offers the ideal-typical example to think about a holiday as a 'black-box'[4] from the perspective of the tourist. That is, the tourist books the package, turns up at the airport and does what they are told from there. Open up the black box, and we can see the work done by the travel agency in drawing

---

3   See Abram and Lien 2011 for a more detailed account, in relation to nature.

4   'Black box' is a term used by engineers for an item in a circuit that will be accepted as producing an effect without its workings being subject to scrutiny – in any explanation only so much can be interrogated at once – some things must remain taken-for-granted for the explanation not to become overwhelmingly complex.

together systems to enable the customer to make a booking, coordinate long and short distance travel, book appropriate hotels, secure food standards, ensure that day-trips are available, and generally ensure that the tourist is steered through the holiday. Yet even the more independent traveller who books their own travel and accommodation, and may spend as much time planning the holiday as taking it, relies on a complex system that generally remains unexplored by the traveller. Booking a flight requires an information system that connects the passenger to the plane, via their bank, and through a recording system that ensures them a seat on the flight they require. Then there is the machine that is the aeroplane, with its multiple interacting systems of control, mobility, communication with flight-control and passengers, and between crew. And the flight requires an airport, itself a complex world of multiple systems, from baggage handlers to security to cleaners, and so on. And behind the airport is a governmental system of regulation and licensing, interacting with chains of private enterprises operating sites, ground-staffing, handling crews.

Infrastructure also includes conceptual frameworks as well as concrete constructions, as Picard makes clear. Associations for hikers, campers, touring groups, etc., may provide the mental framework as well as the equipment necessary to embark on particular kinds of travel. Gro Ween and I (2012) have illustrated the way that the Norwegian Tourist Association created a framework of documentary and physical tools through which the Norwegian hiker can conceptualize the country as a unified landscape that is legible and accessible to movement. Using maps, route-guides, marked paths and various kinds of cabins, the Association offers a way to navigate the landscape with minimal risk and maximum certainty. Through the systematic use of cartography and illustration, it opens up the Norwegian landscape as a space of potential activity, and encourages the tourist to venture into it, and to expand their horizons from one location to the whole country. Looking at what tourists do, what the association does, and how the two interact, we can see how they form an order of tourism that feeds certain kinds of banal nationalism (see Billig 1995) and vice versa.

It is not necessary to go further into detail to emphasize that the kinds of studies we see in tourism research are heavily framed within particular perspectives, and that without such a narrowed focus, the scope of study would be unmanageably extensive. What we have are themed studies: the world of the airport, hotel management issues, traveller experiences, the creation of 'destinations' (see Viken this volume), and so on. Each sheds light on a particular area, leaving the broader whole in the shade. Since that broader shade is modern life itself, an ungraspable abstract, then it is perhaps to be expected. However, when we do look at particular practices, the challenge of the research lies in finding the limits to those practices, deciding how to separate out our research from broader debates in related disciplines, or fathoming out what is particular to tourism, or leisure, as opposed to other activities.

Destination Development as Planned Development

In the case of tourism development, one might argue that there is, in fact, little that distinguishes the core development questions from any other kind of development, since tourism is one of the kinds of development that development planning includes. Planning and building an airport or a hotel are subject to the same regimes of planning and building regulations as a car park or a factory. The aspects of infrastructure that include changes in the use of land, construction of buildings, provision of transport or transport infrastructure, and so on, are likely to be governed (or not) by an existing planning system, and are subject to the same regulations and pressures that apply to infrastructure produced for other reasons, and the critiques and analyses applied in planning studies are relevant for tourism development studies. If we accept that tourism development shares much of its context, its challenges and opportunities with other kinds of development, then tourism development shares many of the governmental challenges that other forms of development do. How should business interests, political ambitions, civic activity and local preferences be managed, for example? How can developments be made sustainable and according to which definitions and priorities?

Planning development is a common activity governed in democratic states by formal bureaucratic processes. It can be argued that state planning is one of the bureaucratic practices that defined the twentieth century, and remains a central axis of state power today in many countries. State planning incorporates the ordering of new developments and the control of existing ones, and at its broadest it is a way of envisaging a future and organizing space through time (see Abram and Weszkalnys 2013). The formal strategic and land-use plans presented by local and national governments offer an intrinsically optimistic vision of a future that can motivate action in the present, at many different scales. Plans are also produced by private corporations as a projection of their anticipated actions, used to generate investment as well as confidence in business. Such plans outline the promise of a future of growth, profits and benefits for certain actors. Benefits may be imagined as shared, most commonly through the creation of employment, which is often the most persuasive of goods promised in development plans to encourage politicians to give their approval. Indeed, the promise of employment and economic growth is often posed as a public good worth sacrificing other benefits for, and hence a schizmogenetic[5] competition between local environmental or social goods and economic growth is commonly the order of the day in planning debates (Murdoch and Abram 2002; Abram 2012). Trapped in this debate, misgivings about development on the part of non-professional participants, such as fears about the diversion of water resources from local smallholdings to luxury hotels, or restrictions of access to beaches, or increased traffic in sensitive ecological districts, can be characterized by adept planning agents as backward looking or

---

5   See Bateson 1973.

anti-capitalist. It is, indeed, difficult to resist development in many contexts, if not impossible (Murdoch and Abram 1998, Abram and Waldren 1998).

One of the central dilemmas in governmental planning, either at central or local levels, is the need for developments to satisfy local conditions while also fulfilling economic goals and fitting into democratic structures. Indeed, one of the key challenges in all democratic development, including tourism development, is to ensure that development is produced in a way that can be considered ethical. Ethical development can arguably be defined as a form that distributes the costs and benefits of development equitably. Since the 1960s at least, many people have argued that one way to achieve this is to ensure that all those potentially affected by development are included in questions of planning and control (see, archetypally, Arnstein 1969). Participating in planning and development is not so simple as it might first appear, however. Participants require considerable technical knowledge to be effective. Arguments must be presented in a fashion appropriate to legal debate, matters of concern must be restricted to those governed by planning legislation, and discussions must be conducted in the context of political debate. This, too, requires considerable technical learning on the part of participants, and those with least assistance are generally the lay population, local residents in the vicinity of proposed developments. Developers, politicians and state planners have professional and paid assistance, while people who happen to live in the vicinity are likely to have less time or resources to fully develop their position (see Murdoch and Abram 2002). For people in poverty, either local residents or minor entrepreneurs and small business holders, the task of taking on a major developer whose interests threaten their own can be overwhelming for different reasons. While the scale of undertaking to represent themselves in a planning process (either in the local political forum or through a legalistic enquiry) is daunting, challenging the mentality of development can be even greater (see Viken in Chapter 2 of this volume).

But development norms can be challenged, and planning systems are constantly subject to change. The activities of multinational campaigning organizations (such as Tourism Concern) play a valuable role supporting non-commercial interests in tourism development as similar organizations do in relation to other forms of development (where national and international environmental charities are particularly active[6]). Their main contribution often lies in helping to legitimize the arguments of non-state actors, by lending the credibility of a national organization and the weight of international research and knowledge, shifting the balance between state agents and private citizens.

Much attention has also been focused on the kind of changes characterized by the globalization of capital and the increasing role of private actors in government processes, often referred to as neo-liberalization. Certainly, freer international

---

6 In Britain, key organizations who often engage in planning debates include the Campaign for the Preservation of Rural England (CPRE), the Royal Society for the Protection of Birds (RSPB), Association of Small Historic Towns and Villages (ASHTAV), Friends of the Earth (FoE), the Victorian Society, and many others.

flows of capital, and easier access to transport for the wealthy are important influences on the international tourism trade. We should remain sceptical about the use of the term neoliberal, however, since its central tenets are not unified into a coherent philosophy, and there is apparently no founding or unifying text. Hence, Peck and Tickell (2002) focus on processes of neoliberalization, while Brenner and Theodore suggest studies of actually existing neoliberalism (2002). Anthropologists, on the other hand, have kept a close eye on the actual elements described as neoliberal, charting the effects of monetarism (Guyer 2007), or a rising tide of managerialism that threatens the place of politics (Ferguson 1990; Abram 2007). Development practices are subject to the same conditions, including the weakening of state regulatory practice, centralization of financial transactions in non-domestic locations, a rise in tax-avoidance practices, the weakening of employment security, managerial strategies that weaken employee participation, and so on. At the same time, a clear feature of development practice during the period of the rise in neoliberalization is the promotion of participative planning. One might argue that this is a paradox, but there are alternative accounts. On the one hand, participative planning and policy-making was a siren-call of the civil rights movements of the 1960s, allied with the growing anti-authoritarianism of the post-war generation. On the other, it is clear that the notion of participative development has been very successfully incorporated into major international institutions such as the World Bank, or global lobbying groups such as the Bertelmann Institute, prime movers in the introduction of neoliberal economic and social policies. Indeed, participative planning has become so well established that, far from generating community-led political consensus, it has been accused of becoming a means to shift the blame for policy failure onto the recipients or objects of policy (Cooke and Kothari 2001).

These critiques leave us with a set of difficult questions about the rights of different parties in development situations. If development should be subject to local democratic control, how is that control best assured? How can a balance of interests be maintained if one actor is significantly over-represented? How can community rights be secured when well-resourced powerful international corporations threaten to overwhelm local political forums? How can new ventures be established when there is not agreement about the political, environmental or social priorities in a region? Given the insecurities that characterize the tourism industry generally (its vulnerabilities to changing fashions, to international economic situations, its dependence on transport links, its reliance on stable international relations, etc.), these questions about development are always present. Rather than expecting a solution at a theoretical level – and as yet, no successful model of participatory democratic policy-making has been proposed – groups generally get stuck in and try to campaign for outcomes they desire. On the one hand, industry actors and investors push for development, on the other hand, activists and pressure groups lobby for consumer rights, environmental rights and community rights at local and international levels (notably Tourism Concern).

One strategy left for academics and theoreticians is to challenge current or accepted theories of tourism development, whether they are found in theoretical debates or in the circulating discourses of tourism and development that help to underpin the range of tourism practices. Here, a combination of philosophical consideration and empirical investigation remains a rewarding approach, and the discussion above of the importance of seeing how things are actually done in practice, how dominant categories and concepts are produced and maintained, through or in spite of discursive rationalities, comes back into focus as a research strategy. Looking at the way that development is done has been central in the growing critique of development practices, both in terms of international aid (e.g. Mosse 2005) and of local development (e.g. Weszkalnys 2010). The whole panoply of socio-cultural analysis is seldom brought to bear on questions of development, yet it offers potentially rewarding and rich insights. I noted above how performance and ritual studies have been incorporated into tourism analysis (through the participation notably of anthropologists in tourism research), yet much less influence is apparent in planning studies. What would be the result of applying ritual and performance theory to participative planning and development? And what might this then offer to the study of destination development? In my view, the results offer insights on the reasons why participatory planning is so often a struggle, since it helps us to see how plans, planners and planning construct and perform ritual practices that exert an exclusive effect on other participants (see Abram 2011).

Where planning for development is a bureaucratic practice (i.e., normally in democratic states), it is practiced according to local rules and rituals that tend to lend legitimacy to presiding authorities. Political debates are structured to enable the voices of elected representatives to be heard, and to some extent to under-emphasize the role of bureaucrats in planning decisions. In Britain, planning officers make recommendations to planning committees, but the committees make the decisions. The effect of the recommendation is noticeable when the two contradict, or where decisions are controversial. Otherwise, planning decisions are often delegated to officers or pass through committees with minimal explicit reference to officer information. Yet the case put to the committee is written and presented by the officers, and thus intrinsically represents their professional view. The routinization of the process helps to underemphasize the politics of the professional's role, while the formal presentation through committee, with its regular agenda, call to meeting, correct behaviours (addressing the chair, taking turns to speak and only on the issue at hand, etc.), use of appropriate language (such as titles, 'Councillor Smith') and clothing, and the performance of the committee's work in the surroundings of the council chamber or relevant committee room, contribute to the legitimizing effect of the planning process. The statement of a decision to allow a development or accept a development plan, for example, is considered valid when delivered by the correct person in the presence of the correct audience, in the right place and documented in the correct way. All these aspects contribute to the ritual aspect of development planning.

Further aspects of ritual theory can also be applied, particularly in relation to the temporal progression of developments. Ritual theorists, notably Van Gennep (1960), but also Turner (1974), note the transformation of both matter and social relations through the performance of ritual action. Van Gennep identified the phases of ritual as, put simply, before, during and after, and noted that during the ritual, the subject of the ritual entered a liminal phase, where they were neither the subject of before the ritual nor the subject of after the ritual. That is, status is suspended during the ritual process, before being re-established as a product of the ritual itself. What does this mean in relation to planning processes? Among the many ways this theory can be considered relevant (e.g. in relation to decision-making processes, in parallel to the analysis above), is in understanding public responses to development projects. In my experience, planners and developers often complain about the difficulty of engaging local communities early in planning and development processes. Most people are unwilling, for whatever reason, to engage in the politics of planning while it remains an abstract discussion. Yet when people see a building site, either as a result of a demolition, or the fencing off of an area ready for development, a sudden surge of interest may emerge. The preparation of a site for development marks the phase of a transformation where the status of the site is uncertain, or liminal. In entering the process of transformation, the reality of its transformation becomes palpable. Demonstrations and public concern that emerges at this stage is often ineffective, since work usually begins on sites after decisions have been made, and decisions having been made, the work is usually invulnerable to further protest. Many of the more violent anti-development protests arise from this dislocation – anti-airport action in Britain by activists occupying or chaining themselves to trees, for example, take physical action when they realize they cannot change decisions made politically. On the other hand, as activists learn how prior decisions have been made, they may well begin to combine public protest with participation in political processes, as has been shown elsewhere (see Murdoch and Abram 2002). Learning the ins and outs of the ritual process, how to intervene in it effectively and when to ignore it, are intrinsic to the career of participants in planning processes, whether that is for housing development, tourism development, or destination development. All this suggests that while we may learn to love tourism, we may also learn to object to tourism development. Some people also learn to dislike tourism, for its environmental impacts or its social consequences, and some people never learn to like being a tourist at all. Without a way to account for cultural variations, we cannot understand how planning disputes arise, or how (or whether) to manage the difficult transition from the present to the future.

Conclusions

My aim in this chapter has been to draw out some of the links between destination development and development planning, to show how theories and analysis from one might be applied to the other, and what might be gained by reflections from different disciplinary perspectives. I have hinted at the possibilities for methodological development, indicated areas of potential value in analytical terms, and argued that the performance of destination development is not merely about the theatricality of tourist products but is intrinsic to the imagination of tourism itself. Acknowledging the philosophical depth to the notion of performance helps us to recognize that we create norms, expectations and routines through everyday actions that might otherwise appear insignificant. This is quite a different route for thinking about performance than the more established mode in tourism research, which adopts the theatre-inspired analysis of presentations, the notions of front and back stage that MacCannell offered as a way to conceptualize the presentation of people and places for tourist consumption (1976).

In raising the spectre of development planning, I might be accused of missing the point of destination development. That is, destination development comprises an ambition to draw together the activities of myriad actors through coordination and combining of forces to present a coherent whole to the tourist-audience. In this, destination development has much in common with what is now called 'town centre management'. Town centre managers attempt to coordinate the services offered in a town centre to make it accessible to the different users who inhabit and visit it. They attempt to take a holistic approach to the town, to keep it safe, attractive, entertaining and enticing. Just as with destination management, the spatial object cannot merely be marketed in a coherent way to tourists if it is not, in itself, coherent. The town centre manager, or the destination manager, needs also to do the work of shaping the place as a manageable and legible unity. Destination development adopts a similar approach, in wishing to create a seamless experience, offering everything a desired sector of tourists might expect. Put in this way, we can see that the construction of tourist desires is then mirrored in tourism development desire, and comes to shape the destination in concrete ways. Destination development is thus also an ordering of desire, performed through a combination of discursive, material and political acts. The performance of destination development happens equally in the political debates about new developments, about policies for tourism promotion, and about public participation in development policy.

Overall, I have suggested that in looking at how we learn to do tourism, at how we create 'tourism' as a field of action and research, we might open up productive ways to understand those practices and, potentially, to change them.

References

Abram, S. (2011), *Culture and Planning* (Farnham: Ashgate).
Abram, S. (2007), 'Participatory Depoliticisation: The Bleeding Heart of Neo-Liberalism', in Neveu (ed.).

Abram, S. and Lien, M.E. (2011), 'Performing Nature at World's Ends', *Ethnos Journal of Anthropology* 76:1, 3–18.

Abram, S. and Waldren, J. (1998), *Anthropological Perspectives on Local Development: Knowledge and Sentiments in Conflict* (London: Routledge).

Abram, S. and Weszkalnys, G. (eds) (2013), *Elusive Promises: Planning in the Contemporary World* (Oxford: Berghahn).

Austin, J.L. (2009 [1962]), *How to do Things with Words* (Oxford: Oxford University Press).

Bateson, G. (1973), *Steps to an Ecology of Mind: Collected Essays in Anthropology, Psychiatry, Evolution, and Epistemology* (St Albans, Australia: Paladin).

Barad, K.M. (2003), 'Posthumanist Performativity: Toward an Understanding of How Matter Comes to Matter', *Signs: Journal of Women in Culture and Society* 28:3, 801–831.

Billig, M. (1995), *Banal Nationalism* (London: Sage).

Brenner, N. and Theodore, N. (2002), 'Cities and the Geographies of "Actually Existing Neoliberalism"', in Brenner and Theodore (eds).

Brenner, N. and Theodore, N. (eds) (2002), *Spaces of Neoliberalization: Urban Restructuring in North America and Western Europe* (Oxford: Blackwell).

Bærenholdt, J.O. et al. (2004) *Performing Tourist Places* (Aldershot: Ashgate).

Callon, M. (1991), 'Techno-Economic Networks and Irreversibility', in Law (ed.).

Cooke, B. and Kothari, U. (2001), *Participation: The New Tyranny?* (London: Zed Books).

Evans, G. (2006), *Educational Failure and Working Class White Children in Britain* (Basingstoke: Palgrave Macmillan).

Ferguson, J. (1990), *The Anti-Politics Machine: 'Development', Depoliticization, and Bureaucratic Power in Lesotho* (Cambridge: Cambridge University Press).

Franklin, A. (2004), 'Tourism as an Ordering. Towards a New Ontology of Tourism', *Tourist Studies* 4:3, 277–301.

Gennep, A.v. (1960), *The Rites of Passage* (London: Routledge & Kegan Paul).

Guyer, J. (2007), 'Prophecy and the Near Future: Thoughts on Macroeconomic, Evangelical, and Punctuated Time', *American Ethnologist* 34:3, 409–421.

Harwood, S. and El-Manstrly, D. (2012), 'The Performativity Turn in Tourism', *University of Edinburgh Business School Working Paper Series* (Edinburgh).

Lave, J. and Wenger, E. (1991), *Situated Learning: Legitimate Peripheral Participation* (Cambridge: Cambridge University Press).

Law, J. (ed.) (1991), *A Sociology of Monsters* (London: Routledge).

Law, J. (2008), 'Actor Network Theory and Material Semiotics', in Turner (ed.).

MacCannell, D. (1976), *The Tourist: A New Theory of the Leisure Class* (London: Macmillan).

—— (1992), *Empty Meeting Grounds: The Tourist Papers* (London: Routledge).

Mitchell, T. (1988), *Colonising Egypt* (Cambridge: Cambridge University Press).

Mosse, D. (2005), *Cultivating Development: An Ethnography of Aid Policy and Practice* (London: Pluto Press).

Murdoch, J. and Abram, S. (2002), *Rationalities of Planning* (London: Ashgate).

Murdoch, J. and Abram, S. (1998), 'Defining the Limits of Community Governance', *Journal of Rural Studies* 14:1, 41–50.

Neveu, C. (ed.) (2007), *Cultures et pratiques participatives: perspectives comparatives* (Paris: l'Harmattan).

Peck, J. and Tickell, A. (2002), 'Neoliberalizing Space', *Antipode* 34:3, 380–404.

Picard, D. (2011), *Tourism, Magic and Modernity: Cultivating the Human Garden* (Oxford: Berghahn).

Turner, V.W. (1974), *Dramas, Fields, and Metaphors. Symbolic Action in Human Society* (Ithaca NY: Cornell University Press).

Turner, B.S. (ed.) (2008), *The New Blackwell Companion to Social Theory* (Oxford: Blackwell).

Urry, J. (1990), *The Tourist Gaze: Leisure and Travel in Contemporary Societies* (London: Sage).

Veijola, S. and Jokinen, E. (1994), 'The Body in Tourism', *Theory Culture and Society* 11:3, 125–152.

Wenger, E. (1998), *Communities of Practice: Learning, Meaning, and Identity* (Cambridge: Cambridge University Press).

Ween, G. and Abram, S. (2012), 'The Norwegian Trekking Association: Trekking as Constituting the Nation', *Landscape Research* 37:2, 155–171.

Weszkalnys, G. (2010), *Berlin, Alexanderplatz: Transforming Place in a Unified Germany* (Oxford: Berghahn).

Chapter 5
A Place for Whom? A Place for What? The Powers of Destinization

Brynhild Granås

Introduction

As a key term in dialogues on tourism, 'destination' is more strongly embedded in political and management discourses than in socio-cultural theoretizations (Framke 2002; Bærenholdt et al. 2004). Still, the idea of a destination is a comprehensive and permeating *raison d'être* of tourism (Cooper et al. 1993, 7), in the way that it helps us assess topics relating to the question of where tourists are 'destined' to go. The idea evokes expectations about an enjoyable voyage for some and economic prosperity for others, and notifies where such imaginations can meet. In this chapter, I suggest an approach to the matter of destinations as an encompassing idea in tourism, which implies defining 'place' as the object of study. I define place theoretically as relational, dynamic and material, in accordance with Doreen Massey's spatial ontology and theory of place. Relational processes where place is performed as a destination for tourists – i.e. where places are destinized – will be suggested as a key question for destination research. The perspective proposed is positioned in a materially embedded social constructivism (Ingold 2000; Massey 2005; Larsen and Haldrup 2006) that emphasizes the relational and performative (Edensor 1998; Franklin 2003; Bærenholdt et al. 2004) production of space and place through corporeal as well as discursive practices (Mik-Meyer and Järvinen 2005). The perspective presented aims to integrate Massey's place theory in tourism studies in a meticulous and premising way, making it a method of analysis (Baldwin 2012; Granås 2012).

Even though tourism researchers sometimes seem to pay scarce attention to the fact, a conceptual definition of 'the destination' is logically required if 'destination' or 'destination development' are to be assessed as the object of study and thus provide conceptual direction for research. If we neglected to define it, the term would only pinpoint the spatial omnipresence of tourism in some form or another in today's world, e.g. in the shape of an idea or the imagination of a potential for tourism development. Even so, to foreground 'destination' conceptually sometimes leads tourism scholars to essential questions about outcomes. For example, Bornhorst et al. (2010) states that we should view the destination as a 'geographical region which contains a sufficiently critical mass or cluster of attractions so as to be capable of providing tourists with visitation experiences that attract them to the destination for tourism purposes ... ' (ibid., 572). Any presence of a 'critical mass' or capabilities to provide experiences, are in this sense more or less preliminary results of previous processes. Thus, to the extent that the question of how any situation or quality comes about is a primary concern, we should pay conceptual attention to this 'how'. The essentialist ontology that clings to 'destination' as a research concept – and thereby also to 'destination development' – makes it a troublesome reference point in social constructivist approaches.

Generally, an important contribution from social science- and other socially oriented perspectives is to illuminate the socially inextricable characteristics of tourism. In some way or another such perspectives aim at exploring how tourism development is conditioned by society and, obversely, how tourism may condition societal development. In order to theorize such complexity, we have to balance between conceptualizing the world too widely – leading us astray in complexity – or too narrowly – leaving us short-sighted and reductive. To put it otherwise; the ability for theory to enable empirical identifications of the touristically relevant needs to be balanced with a capacity to facilitate observations of tourism as enmeshed in social processes. When Framke (2002) identifies the varying and contradictory ways that tourism researchers have defined the destination – as a narrative, as an attraction, as a marketing object, as a place where tourism happens, etc. – he concludes that 'Each approach tends to define the destination in a different way, emphasizing only one facet' (ibid., 93). Hence, Framke observes how a theoretical dialogue on what the destination is tends to differentiate touristic processes conceptually at the expense of holistic perspectives. Bærenholdt et al. (1994) contribute to this differentiation by reserving the concept of 'destination' to signify the imagery, marketing, and symbolism of the tourism industry (ibid., 28), while taking a profound interest in the performance of tourists, understood as the producers of 'tourist places' (ibid., 2).

The perspective outlined in this chapter discusses the topic of destination by verbalizing the term 'destination' and subordinating it to place, providing an integrative perspective capable of doing justice to the empirical integratedness of for example the discrete and strategic aspects of tourism development, i.e. the practices of tourists as well as strategic development actors within tourism. Thus, an important consideration is to avoid construing as separated spheres that which may or may not intersect empirically in the production of space and place. A basic argument here is that Massey's theorizations of space and place make room for such non-differentiating and integrative approaches, for example to the topic of destination.

Throughout, I will be concerned with the question of how to interpret the zone of potentiality implied by 'destination', and how to orient our eye within it by the use of theory, tracing the lingering and sometimes unpredictable trajectories of potentiality that revolve around the tourist destination as a world-embracing idea about travel and about fashioning places where tourists are 'destined' to go. My suggestion is that Massey's theory of space and place can help make sense of the vagueness of destinations (Framke 2002, 92–93), while preparing a more stable conceptual basis for research. I will suggest ways to observe and discuss the conditions for destination development and the conditioning factors it involves in itself. Conditioning elements may appear to be both resources and hindrances. History, nature, or urban landscapes can, or may never, be resourcified in a tourism economy, while regulation regimes or public discourses can block, as well as release, a touristic boom. The chapter presents tools for observing and discussing the when and how about such issues. Further, the possible effects of strategic destination development can be understood as good or bad, depending on whom you ask; in principle, and as an example, both flourishing and poor tourism can be considered a plague among parties with conflicting interests. Hence, the chapter also aims at clarifying power aspects of destination involvement.

Pulling out the Place of the Destination

Even though the term destination may be unstable in the way it addresses many qualities and aspects, it is constant in the way it always implies the touristic qualities of a quarter of a town, of a city or a nation, or of other geographical entities. Geography is there, implicit in the term, but is somehow positioned in the linguistic shadows, where it runs the constant risk of being overlooked and taken-for-granted. When the geography is pulled out, conceptual differentiations between the geography of for example tourists, tourism strategists and inhabitants are easily produced, for example by conceptualizing the destination as something different from the 'tourist place' (Bærenholdt et al. 2004), or geography is reduced to a physical and static location (see Agnew 1987) for tourism to happen upon or within (see e.g. Frisk 1999). The borders of a particular destination may be unclear and contested and may match or mismatch the more or less dynamic borders of identities, economic systems or political and administrative entities. The unclear and fluid about all such delimitations – including the destination and its borders – reflect their socially constructed nature (Urry 1990; Shields 1991; Edensor 1998). No matter this blurredness, the notion of a geographical entity is however always present when the word destination is spoken.

In the following, I will join those who argue in favour of the value for tourism scholars of investigating this geography, and suggest for a certain way of reading it with a clear eye. Where 'tourism' always involves the geographical 'where' (Bærenholdt et al. 2004), 'destination' addresses the 'where' very directly. Saarinen determines the spatiality of destinations through the concept of region (Saarinen 1998; 2004; see also Saarinen in Chapter 3 of this volume). However, in this chapter I will make Massey's place theory (Massey 1994; 2005; 2007) a basis and address place as method of analysis with methodological implications. For the purpose of the chapter's main discussion, I will do it in a way in which place refers to any geographical entity, be it a village, town, region or country. This interpretation and use is indicated by Massey (1994, 154; Allen et al. 1998), but not emphasized in her writings. In this way, the perspective suggested enables that of including the variety of geographical scale that the term destination may empirically refer to (Saarinen 1998, 155).

Massey represents a central and well-known voice in human geography; her ontology of space and theory of place (Massey 1994; 2005; 2007) has influenced the field considerably, and also marked a 'spatial turn' part of social science theory development during the last two decades (Granås 2012; Bjerre and Fabian 2013). Seen via Bourdieu's advice for a reflexive sociology (Bourdieu 1992), Massey takes place out of the zone of the non-reflexive and taken for granted and deconstructs it, before reconstructing it as an analytical concept. As will be explained, the entire operation relies on an ontology of space. While place is theorized and referred to in many ways within the research literature, what distinguishes Massey's contribution is its abstract quality, as reflected in the basic question she poses about what a place is, and thus the potential it provides for being applied as a method of analysis (Baldwin 2012; Granås 2012). In this way, the level of abstraction and scope of the application of her place theory differ from other theorizations. Where Urry – while not distinguishing between 'locality' and 'place' (see e.g. Urry 1992, 71) – has explored how to analyse 'localities' (or 'places') (Urry 1992, 73), Massey takes profound interest in and elaborates thoroughly on what is analysed, i.e. what a place is analysed *as*. Thus, her contribution also differs from those who ask 'What kind of place is this?' or post similar problems that are closer to an empirical level. One example of the latter would be theoretical elaborations that involve phenomenological experiences of place (Relph 1976; Tuan 1977; Viken and Steen Jacobsen 1999).

The 'what kind of place?' question is of particular interest here in the way it requires a theoretical answer in terms of place categorizations or typologies. Places are categorized within non-academic discourses, where for example London is a metropolis, Hawaii is a surfer's paradise or Alta (where I live) is the town of the aurora borealis (*Nordlysbyen* in Norwegian). Among many place typologies, part of the social sciences are those that identify places according to their geographical location, their systemic position or their relation to a certain economic activity or system of production. The latter categorization practice is of particular relevance for destination research. An archetypical example is the 'industrial' place, while another is the 'de-industrialized' place. And a third one is, I would claim, 'the destination'. This statement about destinations is premised on the reasoning above about how the notion of a geographical area always accompanies the term and how 'place' can refer to all such geographical entities, no matter the scale. What I am then pointing at is how we, when we refer to destinations, categorize the place according to its involvement in a particular economy, i.e. a touristic economy. It resembles modernity's industrial place or late modernity's de-industrialized place (see e.g. Edwards and Cotes 1996; Hayter 1997; Butler 2007; Rast 2001), but is linked to a tourism economy deeply rooted in the post-industrial economies of culture, experience, adventure and the like (Lash and Urry 1994; Pine and Gilmore 1999; Gibson and Kong 2005; Bærenholdt and Sundbo 2007). As with that of assessing a place research-wise as industrial, the term destination tends to identify the place through one economic regime, hence ascribing it with one identity only. When destination is made a basis for research, this homogenizing identity ascription is activated. Thus, a risk is already integrated within the research – from the conceptual start – for missing out on the plurality of a place of which tourism is part.

The same risk is integrated when places are categorized as 'tourist places'. Whether intended or not, such identity ascriptions *per se* mark the place as an object for consumption (Urry 1995), and indicate a commodification of place (Philo and Kearns 1993; Ioannides and Debbage 1998). As a tourism scholar, there is no necessary connection between that of holding onto the categorization of place – or space – as 'destination', 'tourist place' (Bærenholdt et al. 2004) or 'tourist space' (Edensor 1998) on the one hand, while disregarding the plural identities of places or spaces studied on the other. On the contrary, for example both Bærenholdt et al. (2004) and Edensor (1998) pursue such pluralities, as did MacCannel when elaborating on the term 'tourist space' (1992, 20). What I am nevertheless thematizing is how suitable and enabling such categorizations are for attending to the plurality of society, space and place. What I propose is that such pluralities can be addressed conceptually in a more directed and consistent way through a more premising application of Massey's place theory in tourism studies.

Reuniting Place and Destination: The Destinization of Place

The approach to the topic of destination through place presented here elaborates on how we can broaden our conceptualization of the geographies of tourism (Ringer 1998, 9) through the use of place as an analytical concept and method. Going through Massey's theory in the following, I will argue that she provides us with an analytical concept of place with broad integrative qualities, which for example can 'reunite' the place and destination that was somehow disintegrated above. The reintegration is accomplished by subordinating the matter of destination under the matter of place. Therefore, I have referred to the place as an alternative study object and the destinizing processes of place as a possible focal topic of interest. Although with similar intentions, this is different from what for example Ringer et al. (1998) do when they conceptualize the destination as a 'critical context', and include the place as one relevant aspect among others for this context (Ringer 1998, 9). With the purpose of discussing the topic of destinations, the task will be to highlight the active role of tourism in the social construction of place through space, rather than of space through place, as suggested by Ateljevic (2010).

The place as relational, material and dynamic

The starting point for Massey's conceptualization of place is a relational and spatial perspective on social reality. She identifies the thread in her own work to be 'the attempt to formulate concepts of space and place in terms of social relations' (Massey 1994, 2). As a result, her spatial theoretical device embraces the basic sociological topic of 'the relational' in the outline of a theory of place. An important basis for her theorizing is the associations the concept of place gives within our culture, tying it to the authentic, the static, the closed and the homogenous (Massey 1994), as is also pointed out by Abram (2011, 12–13). This is how place is also often referred to by social scientists, e.g. by treating places as one-dimensional in terms of identity. With reference to the discussion above about that of defining places according to one identity only, based on economic characteristics, Massey wants to do something about our one-dimensional- and static understanding of place, not least because there is a politics involved that needs to be addressed. From this deconstruction, she starts out her reconstruction of place, reassessing it as an analytical tool.

For what is a place really? To answer this, Massey starts by directing our attention towards space. Space is often understood as part of a dichotomy with time, in which space becomes the static and time the dynamic (Massey 1994, 249). However, in Massey's description, time and space is inseparable. Space is not something abstract and geometrical, as we usually tend to think. Instead, it should be seen as an aspect of the social, i.e. as an aspect of relations in which people, institutions and material surroundings are a part. Consequently, she speaks of space as dynamic (to emphasize this point, she renames space into time-space – see also Thrift 1996). To make space a social concept in this way is a constructivist take that implies that space is socially shaped and reshaped. However, this is not a radical constructivist way of thinking that perceives space to be 'always constructed' (Massey 1992). Neither is space and geography only end products of the social – the spatial about the social is also something that conditions social life. Massey describes this by stating that 'the spatial organisation of society … makes a difference to how it works' (Massey 1994, 254). One example from tourism that can illustrate this point is how a tourism economy is conditioned by close or distant relations to potential markets, which makes a difference in how it can advance.

Space is an aspect of the social and hence accompanies all social relations and social processes. If thinking spatially is new to the reader, it may help to envision an example of a relation. According to Massey, this relation can involve individuals, institutions, as well as the material about our worlds, such as for example clothes, rocks, documents or physical territories. When you have thought of a relation, you may try to identify the location of the parties within it, and then see the space of their relation; understood this way, space is an aspect of social processes that the parties you have identified are involved in, but not statically positioned within. Nonetheless, this envisioning exercise should not lead one to think of the spatial as uncomplicated. When tying space to the social, the spatial is just as intricate as the social itself; it includes imaginations, dreams and expectations, power issues and temporal aspects, along with many other social dimensions that are part of all social relations. Hence, and as Urry (1995) also states, 'there is no simple "space", only different kinds of spaces' (ibid., 66). When space is understood as socially produced (Lefebvre 1991; Simonsen 1996), no general science in terms of for example universal laws about the effects of the spatial can be established (Urry 1995, 66).

In sum, society consists of a multiplicity of space (Massey 1993, 3). In the same way that social processes overlap and intersect – and sometimes also in antagonistic ways – so do spaces. This is why Massey describes space as broken, fragmented and full of paradoxes. According to her, it is within such tensional zones that social life may change: 'within this dynamic simultaneity which is space, phenomena may be placed in relationship to one another in such a way that new social effects are provoked' (Massey 1994, 4). One example of intersections between spatial social processes is described by Gunnarsdóttir and Jóhannesson in Chapter 6 of this volume about tourism development in Strandir, Iceland. Here, a time-specific meeting between a 'public discourse on regional policy, local narratives and ideas about the dynamics, demands and challenges of tourism work, historical accounts of the past and folklore' triggered the development of a tourist destination that was themed according to a history of witchcraft. Not to indicate that Massey's contribution is a theory of innovation, as at this point she nonetheless articulates how – in the case of Strandir by innovating – an outcome of spatial intersections is neither accidental nor inevitable, but possible in the sense of being contingent. Thus, outcomes are describable and can be understood through the spatial processes that intersected and conditioned them.

It is not hard to imagine that the types of connectivities in terms of intersecting processes described above are innumerable in society and how they may intersect any imaginable geographical scale, e.g. between the local and the global. Thereby, Massey's theoretical identification of place through space, as will be explained in the following, attends to the problem of handling societal entities as closed (Bærenholdt and Granås 2008) through conceptualizations of society, community, culture, place and the like (see e.g. Barth 1992; Saugestad 1996; Ingold 2000, 192; Urry 2000, 186). Her particular way of doing it is through a vigorous theoretical outline that emphasizes the relevance of place. At the same time, her outline makes place a method of analysis in the approach to the open, dynamic, material and complex about society, establishing a framework for a socially integrated research approach to the topic of destinations:

Massey sees place exactly as a time-specific coming-together of the spatial social processes described above: 'If, however, the spatial is thought of in the context of space-time and as formed out of social interrelations at all scales, then one view of a place is as a particular articulation of those relations, a particular moment in those networks of social relations and understandings' (Massey 1994, 5). This implies that place is a meeting between social processes that run within relations which are spatially wide-ranging, e.g. so wide-ranging that we would determine them as 'global'. Hence, Massey describes her understanding as 'a global sense of place', further explaining this by picturing the earth as seen from a satellite: From the satellite, let us imagine, she suggests, that we can observe not only all physical movements down there, but also the (partial invisible) economic, political and cultural relations between people. Also, we can pretend that we are able to see communication, experiences and power that cling to such relations (Massey 1994, 154), as it is from this perspective that we can envisage an alternative interpretation of place. Here, the specific about place is not only decided by its history or its natural formations. Rather, part of its uniqueness is the constellation that the place makes up as we move from the satellite towards the place and observe it as a meeting between all these processes:

> Instead, then, of thinking of places as areas with boundaries around, they can be imagined as articulated moments in networks of social relations and understandings, but where a large proportion of those relations, experiences and understandings are constructed on a far larger scale than what we happen to define for that moment as the place itself, whether that be a street, or a region or even a continent. (Massey 1994, 154)

Thus, place is socially constructed but materially conditioned and 'earthed', still without being materially determined. As with anything social, in principle it is always changing and must be understood as time-specific. Place is produced within a wider geography than that which is indicated by any possibly imaginable physical borders. To interpret place and understand it is to see all the relational processes that meet up in a place at a particular moment.

From this position, we can look for spatial relational processes that perform the place as a destination for tourists – that destinize the place. In addition, the wider geography within which the place is produced – for example in terms of destinization – can be identified; what people, institutions, companies or organizations located where on the globe – or in the neighbourhood – make themselves felt within the place we study. And when doing this, we can detect what other social trajectories that traverse the destinizing processes we have already identified. Furthermore, we can describe what occurs in such intersections and how the creative forces of such meetings may nurture or hinder the development of a place towards the idea of being a destination for tourists.

The powers of destinization

Part of what can be investigated through Massey's perspective is power relations that are in play in the meetings described above. Her conceptualization of space and place is outlined according to an ambition of being politically relevant: 'thinking place relationally was designed to (…) intervene in a charged political arena. The aim initially was to combat localist or nationalist claims to place based on eternal essential, and in consequence exclusive, characteristics of belonging: to retain, while reformulating, an appreciation of the specific and the distinctive while refusing the parochial' (Massey 2004, 6). This type of intervention 'in a charged political arena' constitutes an alternative power-critical research perspective in studies on the topic of destination development. Instead of relying on the presupposition of an existing or becoming destination, defined according to a set of characteristics of the destination, a conceptual step can be taken back from the destination without losing sight of the topic of destination. To unveil and settle the place as a study object provides a perspective which enables that of identifying processes within which places are destinized, without prioritizing the touristic about place conceptually. The perspective also allows for that of observing other processes that destinizing trajectories meet up with in place, and then exploring what occurs within such meetings, in terms of for example power-relevant issues of antagonisms, negotiations and creativity. The plurality of place that comes to the surface through this definition of place direct one's awareness towards destinizing processes as one among many identity ascriptions to place. In this way, the identity politics of any destinizing way of performing a place is addressed. The point with such a critical perspective is not to expose any identity political push as oppressive or negative. As Massey says, 'This does not mean that there is no justification for any notion of conservation [of place identity], but it does mean that the debate should focus on the terms and nature of both conservation and innovation' (Massey 1994, 8).

This statement from Massey resembles Michel Foucault (Granås 2012, 31–38) when he states that power is not necessarily 'fraudulent' (Foucault 1982, 786). With a focus on relational processes and how places are performed according to multiple identity schemata, I find it relevant to perspectivate the processes of meaning production that take place in the spatial intersections that Massey identifies with Foucault. In his work, Foucault primarily assesses that power is a matter of productivity, and not of resources. Hence, power is omnipresent and circulating in society, and we are all performing power while at the same time, we are exposed to power (Schaaning 1993, 213). It is this productivity in relation to place that I am thematizing throughout this text, i.e. the societal production of which tourism is part and that we can approach through place as a method of analysis (Granås 2012). Not least, is it relevant to see this when a deeply geographical topic such as destination and an inextricable phenomenon such as tourism are on the research agenda.

In regard to methodology, what I want to direct the attention towards here is the meaning that any practice from any actor – ranging from the large societal institutions to the single individual – may express, no matter the degree of impact on place we imagine it to have. Thus, as the specific geographical topic of destination is put on the agenda, of interest are not only the performances of tourists that may make places 'touristic' (Bærenholdt et al. 2004, 2), but also what inhabitants, politicians, business actors and others may do that imprints on place destinizing processes. Because of this, we can observe both the small steps and great setbacks in destination development. Additionally, both discursive and corporeal practices (the 'talking' and 'walking') are acknowledged as valid sources for exploring the terms and nature of how places are destinized, as well as practices that produce place in relations with material surroundings, as emphasized by Foucault (Neumann 2001) and Massey (see for example Massey 2005, 97–98). Hence, the perspective is not restricted to the discursive or semiological creation of local geographies (see for example Hughes 1999). Instead, it opens up for applying a variety of methodologies that can grasp meaning productive processes to see how they generate types of knowledge and cultural order (O'Farrell 2005, 100) or, to put it otherwise, are involved in formations of epistemological schemes (Pakier and Stråth 2010, 5) that can provide the basis for decisions about the future (Granås 2012). This methodological interpretation and specification exceeds the place theory of Massey, who demonstrates a methodological operationalization of place on several occasions, but does not elaborate explicitly on a methodology of place (Baldwin 2012; Granås 2012). This also makes her theory of place approachable and usable for researchers working from different methodological positions; Massey's perspective makes room for the many constructivist ways of working within the social sciences and humanities in the exploration of meaning productive processes.

Back to the intersection between spatial social relations that comprise place, Massey's advice can enable researchers to simply observe 'what happens' (Foucault 1982, 786) in such meetings. However, while not losing sight of asymmetrical relations in terms of the distribution of resources, we should open up for noting the manifold, complex and sometimes unexpected that happens to place and tourism initiatives, while seeing place as a continuous negotiation:

> What is special about a place is not some romance of a pre-given collective identity or of the eternity of the hills. Rather, what is special about place is precisely that throwntogetherness, the unavoidable challenge of negotiating a here-and-now (itself drawing on a history and a geography of thens and theres); and a negotiation which must take place within and between both human and nonhuman. (Massey 2005, 140)

Through this perspective, any situations that we may identify as researchers are contingent products and preliminary outcomes. Such outcomes cannot be explained through the logic of cause-effect, nor in relation to the history, materiality and nature of place. Even though we tend to think of inert material surroundings such as natural formations, infrastructure or buildings as fixed, we also know that they do change, both due to forces of nature and to human interventions (see for example Kaján and Saarinen in Chapter 11 of this volume, about tourism and climate change in Northern Finland). But just as important here, material surroundings should be understood as something dynamic in that our understanding of them are changeable. To illustrate this point in regard to tourism, the same way that the valuing, practicing and evolvement of arctic landscapes – often described as wilderness – has changed continuously throughout modernity, even places characterized by heavy pollution[1] can be reframed into the appreciable, and then be resourcified in a tourism economy and inscribed into a global patchwork of tourist destinations. Thus, places are materially constructed in processes where the material, the social and the cultural are integrated (Ingold 2000). When Ingold states that ' ... to inhabit the land is to draw it to a particular focus, and in so doing

---

1 'Holiday in Chernobyl? Tips for pollution tourism', *The Guardian*, published online 4 June 2012.

to *constitute* a place' (Ingold 2000, 149), the perspective of this chapter sees any way of relating to the land – by those living there, by those passing through, by those forecasting it from afar, or by those exploring it resource wise – as ways of drawing it to a particular focus, and in so doing, involving in a continuous reconstitution of place.

Concluding Remarks: The Inconvenience of Pluralism?

The production of place meanings addressed through the perspective of this chapter is one that aims at transcending any divisions between the cultural, the political, the economic and the material. Taken together, the interpretation and accommodation of Massey's theory made here can nevertheless be assessed as a cultural one. It resonates with some of the core issues in a cultural studies tradition in the sense of understanding culture as something that 'exists in an unavoidable relation with power' and in the widest sense includes for example political institutions and forms of social life (O'Farrell 2005, 17). The plurality of place that Massey's perspective illuminates is one that bears the risk of being annihilated by the popular development strategies of our times that are aimed exactly at conserving one single place identity, for example through 'place development' as a neo-liberal, symbol- and identity oriented development strategy. Destination development initiatives are among such strategies in the way they are put at work to not only emphasize the place as a tourist destination, but also as a certain kind of tourist destination. Therefore, tourism is often heavily involved in the identity politics that such development strategies imply. Here, strategic actors within tourism may enter into consortia together with other interested parties that work to brand places for the purposes of economic development (see for example Harvey 1989; 2000; Hall 2000). Within such cooperation, not only tourists, but also investors and new inhabitants – preferably of the creative kind so famously predicted by Richard Florida (2004) – are to be tempted through a certain way of identifying a place. Thus, strategic work to develop tourist destinations involve in the production of 'ludic spaces', where places are reduced and homogenized (Lefebvre 1976, 84). Here, the strategists of tourism are partaking in that of turning basic principles of industrial capitalism upside-down, where, 'instead of the circulation of commodities to locations of consumption (places where people lived), post-industrial capitalism has since the 1960s involved the circulation of people to specific locations that are consumed as commodities' (Mitrasinovic 2006, 25). In times when plurality is celebrated as a driver for innovation as well as for appreciated cosmopolitan values, it is at the same time exposed to erasure. What I am addressing as a final comment and implication of the chapter is hence also the relevance for tourism scholars of addressing the topic of pluralism and democracy, and be willing and able to lay bare the 'illusion of a pluralism without antagonism' (Mouffe 2000, 20) also inherent in tourism development as a democratic paradox.

References

Abram, S. (2011), *Culture and Planning* (Aldershot: Ashgate).

Agnew, J. (1987), *Place and Politics* (Boston: Allen and Unwin).

Allen, J., Massey, D. and Cochrane, A. (eds) (1998), *Rethinking the Region. Spaces of Neo-Liberalism* (London: Routledge).

Altern, I. (ed.) (1996), *Lokalsamfunn og lokalsamfunnsforskning i endring* (Tromsø: University of Tromsø).

Ateljevic, I. (2010), 'Circuits of Tourism. Stepping Beyond the 'Production/ Consumption' Dichotomy', *Tourism Geographies* 2:4, 369–388.

Baldwin, J. (2012), 'Putting Massey's Relational Sense of Place to Practice: Labour and the Constitution of Jolly Beach, Antigua, West Indies', *Geografiska Annaler Series B, Human Geography* 94:3, 207–221.

Barth, F. (1992), 'Towards a Greater Naturalism in Conceptualizing Societies', in Kuper (ed.).

Bjerre, H.J. and Fabian, L. (2010), 'Rummet finder sted: Redaktionelt forord', *Slagmark – tidsskrift for idéhistorie* 57, 9–18.

Bærenholdt, J.O. and Sundbo, J. (2007), *Opplevelsesøkonomi. Produktion, forbrug, kultur* (Frederiksberg: Forlaget Samfundslitteratur).

Bærenholdt, J.O. and Granås, B. (2008), 'Places and Mobilities Beyond the Periphery', in Bærenholdt and Granås (eds).

Bærenholdt, J.O. and Granås, B. (eds) (2008), *Mobility and Place. Enacting Northern European Peripheries* (Aldershot: Ashgate).

Bærenholdt, J.O. et al. (2004), *Performing Tourist Places* (Aldershot: Ashgate).

Bornhorst, T., Ritchie, J. R. B. og Sheehan, L. (2010). 'Determinants of tourism success for DMOs and destinations: An empirical examination of stakeholder' perspectives'. Tourism Management, 31: 572–589.

Bourdieu, P. (1992), 'The Practice of Reflexive Sociology', in Bourdieu and Wacquant.

Bourdieu, P. and Wacquant, L.J.D. (1992), *An Invitation to Reflexive Sociology* (Cambridge: Polity Press).

Butler, T. (2007), 'For Gentrification?', *Environment and Planning A* 39:1, 162–181.

Cooper, C., Fletcher, J., Gilbert, D. and Wanhill, S. (1993), *Tourism. Principles and Practice* (London: Pitman).

Edensor, T. (1998), *Tourists at the Taj: Performance and Meaning at a Symbolic Site* (London: Routledge).

Edwards, J.A. and Cotes, J.C.L. (1996), 'Mines and Quarries. Industrial Heritage Tourism', *Annals of Tourism Research* 23:2, 341–363.

Florida, R. (2004), The Rise of the Creative Class: And How It's Transforming Work, Leisure, Community and Everyday Life (New York: Basic Books).

Foucault, M. (1982), 'The Subject and Power', *Critical Inquiry* 8:4, 777–795.

Framke, W. (2002), 'The Destination as a Concept: A Discussion of the Business-Related Perspective versus the Socio-Cultural Approach in Tourism Theory', *Scandinavian Journal of Hospitality and Tourism* 2:2, 92–107.

Franklin, A. (2003), *Tourism. An Introduction* (London: Sage).

Frisk, L. (1999), *Separate Worlds: Attitudes and Values towards Tourism Development and Cooperation among Public Organisations and Private Enterprises in Northern Sweden* (Östersund: Proceedings of Forskarforum: Local och Regional Utveckling).

Gibson, C. and Kong, L. (2005), 'Cultural Economy: A Critical Review', *Progress in Human Geography* 29:5, 541–561.

Granås, B. (2012), *Det gjenstridige mangfoldet. Bak slagord og overskrifter i nordlig byutvikling* (Tromsø: The University of Tromsø).

Hall, T. (2000), *Urban Geography* (London: Routledge).

Harvey, D. (1989), *The Condition of Postmodernity* (Oxford: Blackwell Press).

—— (2000), 'From Managerialism to Entrepreneurialism', in Miles et al. (eds).

Hayter, R. (1997), *The Dynamics of Industrial Location: The Factory, the Firm and the Production System* (Chichester: Wiley).

'Holiday in Chernobyl? Tips for Pollution Tourism', *The Guardian* (published online 4 June 2012) <http://www.guardian.co.uk/travel/shortcuts/2012/jun/04/holiday-chernobyl-tips-pollution-tourism> (home page), accessed 20 June 2013.

Hughes, G. (1999), 'The Semiological Realization of Space', in Ringer (ed.).

Ingold, T. (2000), *The Perception of the Environment. Essays on Livelihood, Dwelling and Skill* (London: Routledge).

Ioannides, D. and Debbage, K.G. (eds) (1998), *The Economic Geography of the Tourist Industry* (London: Routledge).

Järvinen, M. and Mik-Meyer, N. (eds) (2005), *Kvalitative metoder i et interaktionistisk perspektiv. Interview, observationer og dokumenter* (København: Hans Reitzels Forlag).

Kuper, A. (ed.) (1992), *Conceptualizing Society* (London: Routledge).

Larsen, J. and Haldrup, M. (2006), *Following Flows. Geographies of Tourist Performances* (Roskilde: Roskilde Universitet).

Lash, S. and Urry, J. (1994), *Economies of Signs and Space* (London: Sage Publications).

Lefebvre, H. (1976), *The Survival of Capitalism* (London: Allen and Unwin).

—— (1991), *The Production of Space* (Oxford: Blackwell).

Massey, D. (1992), 'Politics and Space/Time', *New Left Review* 196, 65–84.

—— (1994), *Space, Place and Gender* (Cambridge: Polity Press).

—— (2005), *For Space* (London: Sage).

—— (2007), *World City* (Cambridge: Polity Press).

Mouffe, C. (2000), *The Democratic Paradox* (London: Verso).

Mik-Meyer, N. and Järvinen, M. (2005), 'Indledning: Kvalitative metoder i et ineraktionistisk perspektiv', in Järvinen and Mik-Meyer (eds).

Miles, M. et al. (eds) (2001), *The City Cultures Reader* (London: Routledge).

Mitrasinovic, M. (2006), *Total Landscape, Theme Parks, Public Space* (Aldershot: Ashgate).

Neumann, I.B. (2001), *Mening, materialitet, makt. En innføring i diskursanalyse* (Bergen: Fagbokforlaget).

O'Farrell, C. (2005), *Michel Foucault* (London: Sage Publications).

Pakier, M. and Stråth, B. (2010), *A European Memory? Contested Histories and Politics of Remembrance* (Oxford: Berghahn Books).

Philo, C. and Kearns, G. (1993), *Selling Places. The City as Cultural Capital, Past and Present* (Oxford: Pergamon Press).

Pine, B.J. and Gilmore, J.H. (1999), *The Experience Economy* (Boston: Harvard Business School).

Rast, J. (2001), 'Manufacturing Industrial Decline. The Politics of Economic Change in Chicago, 1955–1998', *Journal of Urban Affairs* 23:2, 175–190.

Relph, E. (1976), *Place and Placelessness* (London: Pion Limited).

Ringer, G. (1998), *Destinations: Cultural Landscapes of Tourism* (London: Routledge).

Saarinen, J. (1998), 'The Social Construction of Tourist Destinations. The Processes of Transformation of the Saariselkä Tourism Region in Finnish Lapland', in Ringer (ed.).

—— (2004), 'Destinations in Change. The Transformation Process of Tourist Destinations', *Tourist Studies* 4:2, 161–179.

Saraniemi, S. and Kylänen, M. (2011), 'Problematizing the Concept of Tourism Destination: An Analysis of Different Theoretical Approaches', *Journal of Travel Research* 50:2, 133–143.

Saugestad, S. (1996), 'Mellom modeller og virkelighet. Lokalsamfunnsforskning i Tromsø, Norge og verden', i Altern. (ed.).

Schaanning, E. (1993), *Kommunikative maktstrategier. Rapporter fra et tårn* (Oslo: Spartacus forlag).

Shields, R. (1991), *Places on the Margin. Alternative Geographies of Modernity* (London: Routledge).

Simonsen, K. (1996), 'What Kind of Space in What Kind of Social Theory?', *Progress in Human Geography* 20:4, 494–512.

Steen Jacobsen, J.K. and Viken, A. (eds) (1999), *Turisme. Stedet i en bevegelig verden* (Oslo: Universitetsforlaget).

Thrift, N. (1996), *Spatial Formations* (London: Sage).

Tuan, Y.F. (1977), *Space and Place. The Perspective of Experience* (London: Edward Arnold).

Urry, J. (1995), *Consuming Places* (London: Routledge).

—— (2000), 'Mobile Sociology', *British Journal of Sociology* 51:1, 185–203.

—— (2002), *The Tourist Gaze. Leisure and Travel in Contemporary Societies* (London: Sage).

Viken, A. and Steen Jacobsen, J.K. (1999), 'Forestillinger om steder', in Steen Jacobsen and Viken (eds).

PART II
Catalysing Themed Destinations

Chapter 6

Weaving with Witchcraft: Tourism and Entrepreneurship in Strandir, Iceland

Guðrún Þóra Gunnarsdóttir and Gunnar Thór Jóhannesson

Introduction

Rural areas in Iceland have gone through vast economic and social changes in recent decades. Tourism has increasingly been promoted as a viable option to boost local economies with the effect that the countryside is, now more than ever before, being bought and sold as an experience, packaged and marketed to prospective tourists. The Strandir region, a scarcely populated coastal area in northwest Iceland, has steadily taken on a touristic 'flavour' in recent years. Traditionally, the economic backbone has been sheep farming and coastal fisheries, but both of those sectors have been in decline for several decades. In Iceland the image of isolation, magic, and strong-willed characters is associated with the region, and those characteristics have become important resources in its emergent tourism economy. The recent turn to tourism as an economic option has been facilitated by the Museum of Icelandic Sorcery and Witchcraft, established in 2000. The Museum brings together, and re-awakens, the seventeenth-century period of witchcraft, during which Strandir was one of the most notorious regions in Iceland for witch hunts and burnings. Being one of the first examples of the active use of cultural heritage for tourism development in Iceland, the Museum has played a central role in weaving Strandir together as a potential tourism destination.

The aim of this chapter is to explore the dynamics of tourism development in the Strandir region through accounts of the establishment of the Museum of Icelandic Sorcery and Witchcraft and its role as catalyst for further enhancement of tourism.

At centre stage is a story of three local entrepreneurs who played a key role in creating the Museum. However, the narrative of the Museum's establishment and its role in the emergent tourism economy in Strandir is also related to other accounts. Those are, among others, public discourse on regional policy, local narratives and ideas about the dynamics, demands and challenges of tourism work, historical accounts of the past and folklore. It is argued that the establishment of the Museum of Icelandic Sorcery and Witchcraft can be described as partly improvisational process of 'weaving the world' (Ingold 2000) that relates diverse narratives in time and space. It is highlighted that the success and wider impact of the Museum as a catalyst for tourism development is primarily based on a high level of personal engagement, 'entrepreneurial enthusiasm' and improvisation.

The chapter is based on data collected during four short field visits (5–10 days each) to Strandir in the summers of 2011 and 2012. Semi-structured interviews were conducted with all of the tourism entrepreneurs in Strandir who operated businesses during the period July 2011- July 2012, as well as municipality leaders and rural development advisors. The interviews focused on tracing the process of tourism development in Strandir, as well as identifying dominant practices, issues of concern and ideas regarding future development. Particular attention was paid to tourism resources and the image of Strandir as a destination. The interviews were conducted as part of an extended study of the making of Strandir as a tourism destination that also included informal discussions with tour operators, other local inhabitants and tourists. In total, 22 interviews were recorded and transcribed before being thematically coded and analysed.

The chapter starts with the discussion of tourism entrepreneurship, with a special focus on lifestyle entrepreneurship. We then present the Strandir region before we move on to describe intersecting narratives of tourism development that illustrate the dynamics of the emergent tourism economies in the region.

Tourism Entrepreneurship and Creativity

Rural communities all over the world have been struggling in the wake of socio-economic changes manifested in declining significance of primary industries and depopulation (Bærenholdt 2011). Tourism has been identified by transnational institutes and national authorities as providing an important new source of income and employment, as well as fulfilling the broader role of breaking down social isolation and encouraging the re-population of rural communities (OECD 1994). It is important to note that inhabitants in rural areas are not passive victims of globalization and market demands stemming from external (urban) forces. People have actively taken part in coping with societal changes through various tactics, thus shaping the contours of places in different and creative ways (Bærenholdt and Granås 2008; Nyseth and Viken 2009). In this process, tourism entrepreneurship has been a prominent agent of change (Lane 2009).

Entrepreneurship is commonly defined as the ability to recognize opportunities for change and the willingness to exploit them, which often requires a capacity for risk-taking (Ateljevic and Li 2009; Jóhannesson 2012). Entrepreneurship and innovation are two of the staples in regional policies intended to boost local economies. Furthermore, both are linked to a recent buzzword in regional development policies, namely creativity (Florida 2004; Richards 2011). Creativity is often related to the production of novelty and is closely tied with innovation in the sense of 'radical disjuncture' (Ingold and Hallam 2007, 2). Another way to approach creative capacities is through the concept of improvisation that stresses the relational aspects of creative processes and the way in which people 'construct culture as they go along and as they respond to life's contingencies' (Bruner 1993, 326, in Ingold and Hallam 2007, 2). This is a type of creativity that does not adhere to a 'script' or a detailed cognitive plan but grasps how, for instance, social orders are maintained or continued in the face of constant movement of the world or societal changes.

Here, we follow a similarly relational approach to entrepreneurship underlining that entrepreneurship, as a cultural practice, is a form of relational ordering (de Laet and Mol 2000; Jóhannesson 2012). From this perspective, innovation is also a relational effect accomplished through improvisational net-*work* of 'hybrid collectives' (Callon 2004). It is not as though humans are ordering and re-ordering passive objects, be it artefacts of cultural or natural heritage, on a blank surface of the earth according to a preconceived constant blueprint (Gren and Huijbens 2012). Rather, we are constantly in the midst of things, working through and in-between the spheres we use to identify nature and culture (McLean 2009) – 'weaving the world' through our relational practices (Ingold 2000). Those categories are thereby seen as products or effects of an ordering process instead of a starting point. As such, entrepreneurial activities manifest efforts of weaving the world, making particular events and realities more probable than others (Bauman 2001).

When it comes to the realities of rural tourism a common characteristic is the fact that most tourism firms are small, owner operated and family run with many fitting the category of lifestyle entrepreneurship (Getz and Carlsen 2000; Shaw and Williams 2002). Lifestyle entrepreneurs have not entered the sector solely because of perceived economic gain, and it has thus been argued that they may hinder the growth of 'real' entrepreneurs; that is, those that adhere to a capitalistic growth paradigm (Shaw and Williams 2002; Ionnides and Petersen 2003). Studies have shown, for instance, that lifestyle entrepreneurs are often reluctant to cooperate; their involvement with the formal side of the industry sector (such as tourism boards) is often limited (Peters et al. 2009), and their will to become involved in systematic approaches to sustainable development may be restricted (Haven-Tang and Jones 2012).

On the other hand, it has been clearly demonstrated that the lifestyle entrepreneurs are of great importance for tourism development, not the least in rural areas. These entrepreneurs are often pioneers in introducing new services to areas where more market and growth-driven entrepreneurs are hesitant to act. Lifestyle entrepreneurs are often closely connected to the community where they choose to run their businesses. In this regard, Bosworth and Farrell talk about the embedded tourism entrepreneur as someone who uses his or her local connection and knowledge as a way to open up access for the tourists to the local community, thereby providing them with a more authentic experience and allowing for creative consumption by the tourists (2011). This may further assist tourism providers in terms of carving out a business advantage within an increasingly homogenous marketplace of cultural tourism (Richards and Wilson 2006).

Realizing the potential of tourism can be a demanding task. In scarcely populated rural communities, frequently only few individuals are active tourism players who may feel that they do not have the capacity to do more. Local entrepreneurs thus sometimes find themselves in a difficult situation: on the one hand, they are encouraged to 'set off', but on the other, they find their situation vulnerable because of limited support and resources.

In this regard it is well recognized that co-operation and community involvement are crucial for the development of a rural tourism region (Mitchell and Hall 2005; George et al. 2009). Research indicates that small businesses in rural areas must work in collaboration with each other in order to gain for example stronger marketing power (Cawley et al. 2002; Roberts and Hall 2004). In fact, Ateljevic argues that the tourism sector is exemplified by 'its ability to stimulate broad cooperation and a collective entrepreneurship in a given locality' (Ateljevic 2009, 170). In this regard, none governmental organizations (NGOs) have been an important factor in regional development, often representing a bottom-up development (Ateljevic 2009). The so-called institutional entrepreneurship is characterized by a bottom-up development involving actors from both the private and the public sector who, by pulling together resources, establish new institutions or transform older ones (Ateljevic and Li 2009). Nevertheless, institutional entrepreneurship invariably requires the engaged individual/s who has the power to mobilize different actors through inspirational and creative acts (Huijbens et al. 2009).

To sum up, rural tourism development is faced with the fact that many rural communities are strained by continued depopulation and the dominance of small and micro businesses that may lack the motivation and capacity for product development and provision of quality services. However, the small-scale lifestyle entrepreneurs may be the only viable tourism actors available and may potentially become ever more important for tourism development as they can offer creative spaces of consumption and production thus connecting to the increasingly prominent trend of creative tourism economies (Richards and Wilson 2006; Richards 2011). We will now move on towards exploring how this situation plays out in a particular case in Iceland, the Strandir region.

Strandir

Geographically, Strandir belongs to the Westfjord region that is a peninsula carved and cut by long fjords connected with often-difficult mountain passes. On the map, the region seems to be stretching to the West away from the mainland (Figure 6.1). The Westfjord region is very scarcely populated (about 7,000 inhabitants in an 8,900 km^2 area) and it has experienced more outmigration than any other region in Iceland.

The population of Strandir is about 670 inhabitants split into three municipalities (Statistics Iceland, n.d.), with the main population centre being Hólmavík with 391 inhabitants. Despite the smallness of the town, the level of service is fairly good and various administrative, social and cultural institutions are located in Hólmavík. Furthermore, Hólmavík is the only spot in Strandir that has a good road connection to the rest of the country. The road north of Hólmavík, however, is still in rather poor condition. The northernmost municipality, Árneshreppur, is rarely accessible by road during the winter time, and the rest of the year the bumpy

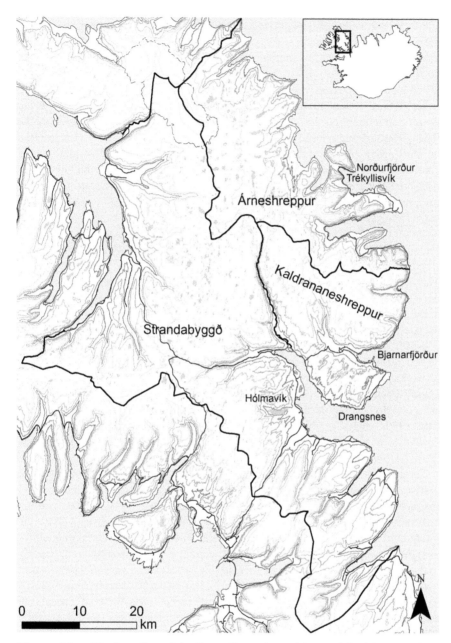

Figure 6.1 **The location of the Strandir region with municipal boundaries and main roads**

Source: Friðþór Sófus Sigurmundsson 2013

and twisted gravel road is regarded as a hindrance for many who seek to travel to the northern part of Strandir.

The region of Strandir experienced a rapid boom in the 1930s and 1940s when herring fisheries blossomed in the area (Sigurðsson et al. 2007). This period only lasted for about 20 years when the herring stock moved to other grounds. However, fisheries have continued to be the backbone of the economy together with sheep farming although both have been subject to intense re-organization, such as the introduction of Individual Transferable Quota in the fisheries and strict regulation on meat production. From the middle of the twentieth century, outmigration has continued until just recently. In fact, a slight increase in the population can be measured since 2005 and lack of housing is now considered to be serious barrier for a further increase (Þorgrímsdóttir 2012).

Icelandic Regional Policy and Tourism

The first account that we would like to connect to in our description of tourism development in Strandir has to do with the link between Icelandic regional policy and tourism. Tourism in Iceland has increased steadily over the last two decades and now plays a significant role in the Icelandic economy, generating one-fifth of the nation's foreign exchange revenues (Statistics Iceland 2011). Restructuring and changes in fisheries management that took place in the 1980s as well as regulatory changes in sheep farming have affected rural communities greatly (Benediktsson 2001; Karlsdóttir 2008). Tourism has been on the radar as a tool for economic diversification in those communities at least from the early 1990s when the public authorities started to attend more seriously to tourism development and perceive it to be an option to diversify and boost local economies (Byggðastofnun 1999; Jóhannesson and Huijbens 2013).

Two issues regarding regional policy deserve special attention. First, cultural tourism has, in particular, been identified as a feasible tourism development option for rural areas. This emphasis by central authorities came very clear around the millennium (Byggðastofnun 1999; Samgönguráðuneytið 2001). As such, it can be interpreted as both a manifestation of an international trend towards 'economies of signs' (Lash and Urry 1994) or cultural economies and a need felt domestically to diversify the tourism supply of Iceland that is still based very much on nature tourism (Gunnarsdóttir 2011; Karlsson 2011). During the last decade, this emphasis has further taken the form of so-called cultural contracts between the central authorities and regions (in a general coalition of municipalities) (see for example Ministry of Industry and Commerce 2005). The aim of the cultural contracts is specifically to enhance cultural activities within the regions as well as support cultural tourism development.

Second, at the same time, regional policy in Iceland started increasingly to emphasize innovation and entrepreneurship. The idea has been to step back from a centralized top-down approach towards a more bottom-up policy where the emphasis is on 'boosting local initiative, enterprise and ideas' (Byggðastofnun 1999, 4). In line with that, certain growth agreements have been made with regions around the country, seven in total. It is then each region's responsibility to employ those agreements to facilitate projects that reflect the strength and characteristics of each region. All seven growth agreements have identified tourism as one of their main focus areas although the success in translating regional potentialities into a positive tourist experience has varied.

Tourism Dynamics in Strandir

In order to follow some of the dynamics of tourism development in Strandir we now turn to other accounts of tourism activities in the region. We start by tracing the process of tourism in Strandir, then we turn our attention to the initiation of cultural tourism in the region, and finally we meet the key actors behind the Museum of Icelandic Sorcery and Witchcraft.

Tourism Connections

The tourism landscape in Strandir is characterized by small local enterprises, just over twenty in total. Some operate their businesses only during the summertime and many have multiple operations such as accommodation, restaurant and a camping ground. The total number of tourists in the region is not known but in the year 2012 around 10,000 tourists visited the tourism information centre in Hólmavík, just about 70% of those were foreign tourists. In 2005, the total number of overnights in the region was 13,715 compared to 18,658 in 2012 of which 57% were by foreign tourists (Statistics Iceland 2013).

Hotels and guesthouses have been operated in the region for many years. In our interviews, most of the accommodation providers stated that they started their business because they wanted to make better use of their properties, and tourism was in fact often described as some kind of a happening or coincidence: 'I took a shot in the dark and was actually just a bit surprised when the foreigners started to come' (interviewee who started a guesthouse in the 1990s). Another entrepreneur was a visitor in the region and felt that there was a need for a place that offered proper coffee. Then there are those who were asked by the local authorities to keep places in operation:

> this just started because we were asked. It was not because we really wanted to enter the tourism business. (Operator of a guesthouse during summertime)

This manifests the vulnerability of rural tourism businesses as a restaurateur in Strandir notes:

> Once somebody leaves and closes the shop, the guesthouse or the café will
> collapse along with it.

Opportunities are clearly identified and ideas for product development are
abundant. However, the question: 'Who is going to do it?' is not so easily answered
and informants repeatedly mentioned the intensity of the summer season with its
heavy workload that to many came as a surprise when they were starting their
tourism businesses:

> When I look back I didn't imagine being this tied down during the summer.
> You know, it just can't occur to me to just take off; go away this weekend.
> (Guesthouse owner)

On the other hand, the tourism actors we interviewed felt energized by positive
feedback and the fact that their efforts are recognized by tourists from all over the
world. As one of the municipality leaders stated:

> to receive tourists, that is just psychologically good. ... [tourism] gives always a
> lot of hope of unutilized opportunities. I think it is a line of business that ... like
> tickles people to take off. Imagination that allows them to do some new things.

This rendering aptly describes how tourism, and cultural tourism in particular,
came to be seen and enacted as an alternative in the Strandir region.

Initiating Cultural Tourism in Strandir

One of the earliest examples of a cultural tourism development project is the
establishment of the Museum of Icelandic Sorcery and Witchcraft in the Strandir
region. It originates in a report written for the county council of the Strandir region
titled *Tourism and National Culture* (Jónsson 1996). Its author was an ethnologist
who had been brought up in the region. The report is one of the first attempts
to relate discourse on tourism development and national heritage and culture in
Iceland. As mentioned above, this emphasis was to be taken up by the central
authorities and has been increasingly prominent over the last ten years.

When interviewed about the idea behind the report, the author, Jón Jónsson
described how he had always wanted to be able to relate his education with what
he labels as a more practical component. Thus, before he contacted the county
council, he had been involved in various museum projects around the country.
Armed with the experience from those projects and the longing to do something
for his home region, he proposed to the county council of Strandir to explore the
historical characteristics of Strandir and the ways in which they could be utilized in
tourism. This met nicely with the concerns of people in the region that something
had to be done to revive and boost local economy and culture in the face of the
continuous struggles of agriculture and fisheries.

Jónsson's report stresses the potential role of small-scale tourism in the region. It details many ideas of possible projects such as exhibitions on the history of the region, one being the possibility to present the history of sorcery and witchcraft in the area. This was a short paragraph and only one item of many. However, it became the inspiration for a creative partnership between engaged individuals and the local municipality that led to the establishment of the Museum of Icelandic Sorcery and Witchcraft that opened in 2000.

In basic terms the Museum is a gathering of historical facts about the period of witch hunts and burnings in the country during the seventeenth century coupled with a collection of Icelandic folklore. The idea of a museum of witchcraft and sorcery was generally well met among the inhabitants although some were sceptical about using this darkest part of the region's history as a springboard for developing tourism.

The local authorities supported the project in the beginning and joined the group in the establishment of Strandagaldur (the name literally means 'the magic of Strandir'), a foundation that owns and runs the museum. Strandagaldur is rooted in concerns about the future development of Strandir, and the original idea was that the museum would be split up in exhibitions in different locations distributed throughout the region and thus firmly linking the whole region together as a one destination woven together with the theme of witchcraft. The museum is now operated in two locations, firstly the exhibition in Hólmavík (opened 2000) and then secondly the Sorcerer's Cottage (opened 2005) in a fjord north from Hólmavík. Plans are ready for the third location in the northern most part of Strandir but financing has been difficult in the aftermath of the financial crisis in Iceland in 2008.

Entrepreneurial Tactics

Leadership is important to rural tourism development and also for encouraging a broad community involvement in the process (George et al. 2009). This is certainly reflected in the story of Strandagaldur where certain individuals are usually identified as instrumental in carrying out and implementing the original ideas and vision behind the witchcraft project. The role of Jón Jónsson can in many ways be described in the traditional terms of the visionary entrepreneur who carries out his ideas through his or her skills, perception of opportunities and willingness to take risk (de Laet and Mol 2000). However, he did not act alone or according to a master plan. In interviews, Jón[1] placed emphasis on the importance of being able to recruit people he trusted and who were ready to dive into the project, so to speak. Amongst his closest collaborators were Sigurður Atlason (Siggi) who has been the director of the museum since its opening and Magnús Rafnsson, a historian, who was instrumental in gathering and organizing the narrative contents of the display. Both were living in the area at the time. According to the small

---

1   It is common in Iceland to refer to individuals by first name instead of by title and last name. We use that convention here.

group, it was partly a coincidence that they chose to start working with the heritage of magic and witchcraft, the main reason being simply that they found this period of Iceland's history interesting.

Siggi and Jón had worked together in the local theatre group, which in Jón's opinion is the best breeding ground for individuals who are willing to engage in new projects and make something happen. Those are valuable qualities in a sparsely populated community where most individuals have multiple tasks and responsibilities. The theatre world is also prominent in the operation of the museum and Siggi continues to regularly perform magic rituals for visitors. In fact his persona is so closely connected to the exhibition that he has got a nickname from it and is usually called Siggi Galdur (Siggi Magic). Siggi's performances are, however, no incidental jokes on his behalf; they are strategic acts in order to use all available means to secure the source of income by creating more spending opportunities for the tourists while visiting the museum. Thus, along with the museum, Strandagaldur runs an online shop, a small café and day tours. Strandagaldur has also made a contract with the municipality to run a tourism information centre beside the museum, thereby securing more stability to the business.

Despite the importance of financial incentive, Siggi's entrepreneurship is also guided by other principles. Thus he stresses that an important motivation for running the café is the fact that no restaurant is operated during the off season in Hólmavík and the day tours are an effort to offer more diverse activities in the area. As Siggi explains: 'Strandagaldur has always seen itself as a leader, if you will, to pull the wagon in this and that [direction]'.

Going back to the development of the museum, both Siggi and Jón stressed the importance of respecting the views of those who voiced their concerns about digging into a dark period of the Icelandic nation and Strandir region in particular. The theme of magic and witchcraft was indeed not established in a vacuum since through the centuries people in Iceland have associated the communities in Strandir with witchcraft and sorcery (Rafnsson 2003). The beginning of a period of witch hunts and burnings commenced in the region with the first three individuals being burned to death in Trékyllisvík, in the very vicinity of the museum (Figure 6.1). Siggi described the relief amongst the local guests at the grand opening of the museum when they discovered that the historical accounts were treated with care and dignity through the museum's exhibition.

This signals an important point in the running of the project, which is that from the very beginning, Strandagaldur has highlighted the importance of basing its projects and exhibitions on sound knowledge and scholarship. Hence, the involvement of the historian, Magnús Rafnsson, has been important and from the very beginning, he has been actively engaged in researching material related to the period of witchcraft and sorcery in Strandir. Strandagaldur has also started or participated in other projects intended to boost the knowledge base of the Museum. It established the Icelandic Centre for Ethnology and Folklore and has collaborated with the Natural History Institute of the Westfjords as well as the University of Iceland in projects focussing on the Middle Ages (see e.g. Edwardsson and

Rafnsson 2006). Hence, Strandagaldur has encouraged the establishment of new partnerships, thereby creating an important connection for Strandir to practices and methods of academic research and scholarship.

Creative Turns – Improvised Orderings

The story of the establishment of the museum and the engaged individuals who made it happen has an aura of a romantic success story of few creative persons but at closer look it becomes clear that the entrepreneurial endeavours or tactics are accomplished through relational work that Callon (2004) refers to as hybrid collectives. Hybrid collectives underline that entrepreneurship and innovation is accomplished through distributed agency, improvisation and relational ordering. The human collective of the three individuals has been dependent on various elements in order to set into motion the Museum of Icelandic Sorcery and Witchcraft and to define and construct the heritage of magic in Strandir as an economic resource.

To begin with, without doubt, the narratives of magic lingering in the cultural landscape of the region afford a basic connection to a potential tourism product in the form of the Museum. The characteristics of the region are the main thread in Jón's report and several of our interviewees mentioned how people in Strandir have through the centuries been shaped and moulded by their close interaction with nature and a harsh environment. This has created numerous survival stories where witchcraft of some kind often plays a crucial role. The image of being apart, remote and historically close to ideas of witchcraft, magic and uncontrollable nature has been enacted in creative ways. The application of magic is in itself a way to control one's situation – to find ones way in an uncertain world and to fulfil one's will despite unfavourable situations (Tulinius 2008). In that way, the entrepreneur and the sorcerer are both responding to their environment with the wish to change their situation or to re-order the stakes of their lives.

Furthermore, although Jón, Siggi and Magnús were central figures in the establishment of the Museum, the ideas underlying the work were not necessarily theirs. Jón, for instance stressed that the report was a gathering of ideas from the people in the region, not his own constructs. This is both important in regard to unpacking the solid role of heroic entrepreneurs and in regard to engaging and enrolling local people into the project to take part in weaving the region, witchcraft and tourism together. Thus, the project did not come about as a play put up on a passive stage but rather as an effect of an improvised interplay between earthly substances and narratives from the past as well as discourses and material landscapes of the present (see also Lund and Jóhannesson, under review). This constant moulding and interactions of different actors also echoes the view of many of our interviewees who frequently, as mentioned above, emphasized the connections to tourism mobilities. Indeed, many saw tourism as a venue for creative encounters. They felt that they themselves were an integral component of the tourism product not only as individuals but also as people belonging to a community shaped by its peripheral location. Existence in this periphery becomes an endless source of stories for those who are interested and ready to listen as reflected in the following comments by our interviewees.

> ... we are just real people and we are part of that, we are Icelanders and ... and we are people who hopefully other people enjoy meeting. (Guesthouse owner)

> There is an interesting sentence that we came across somewhere. The tourist sees the environment and if he travels slowly he is able to experience the environment and get information about the history and culture. The local person, however, she is living the history and is the culture and it is like, to get those two to meet ... And then ... the tourist will be one more story in the story collection. (Tour operator and former hotel manager)

Finally, Jón and the other actors were able to connect to an increasingly prominent discourse about cultural tourism in the early years of the project. The Icelandic Ministry of Tourism at the time showed a great interest in cultural tourism in various ways and the project got significant financial support from the financial committee of the Parliament its first year, which was crucial for its establishment. Since then, the museum has often received grants from public bodies and has received status in public discourse as a showcase example on how to harness cultural heritage in tourism development. This has become an important factor, both for the credibility of the project and it has also enhanced the status of the key actors involved. The project has been supported by the county council as well as municipal authorities in the region through the years, and this has enabled the expression of local solidarity towards public bodies on a national level. It has been demonstrated that entrepreneurial reputation can be instrumental in destination development (Strobl and Peters 2013) and that is to some extent reflected in the case of Jón and Siggi who many of our interviewees refer to as role models and important instigators of new ideas.

Catalyst for Tourism?

The accounts of the Museum of Icelandic Sorcery and Witchcraft and the Strandagaldur foundation bring forth some of the characteristics of lifestyle entrepreneurship. Both ventures are based on a personal engagement and other motives than an economic growth paradigm, although being sustainable in economic terms is of course an important goal. At current there is no direct financial support to the Museum from the municipalities or the regional development bodies. The negative aspects commonly associated with lifestyle entrepreneurship such as lack of skills and cooperation are not substantiated in this case. It is also apparent that the usual profile of the lifestyle entrepreneur as someone working more or less alone does not fully grasp the activities described here. The Strandagaldur foundation has become a collective endeavour with important institutional backing; however, it is difficult to estimate its success. The economic impact of the Museum remains mostly in the fact that it has enhanced the image of the Strandir region, placed it on the map so to speak, and it is becoming an increasingly popular attraction with the effect of stopping tourists in Hólmavík where the main exhibition is. This has without a doubt strengthened the base for service in the area.

The social impact of the Museum has probably been much greater than the economic impact. The project has proven that it is possible to develop a tourist attraction in a peripheral region like Strandir and has thus served as a catalyst for alternative efforts of economic development. Many tourism entrepreneurs that we interviewed also mentioned the importance of being able to seek both advice and encouragement from the museum instigators: "… and they, of course, both Jón and Siggi have been like fountains of ideas", according to an operator of a cultural exhibition. While the original plan to create a witchcraft and sorcery trail through the region has not quite been realized, there was a strong consensus among our interviewees that the Museum has been central in tourism development in Strandir. A feeling of scepticism that some had in the beginning has now been replaced by a feeling of pride and self-empowerment or as one entrepreneur stated 'and then to become proud of one's reputation is really one of the things that the museum has done'.

The Strandagaldur foundation and not least the Museum itself have served as a role model when it comes to professionalism and quality in service and products for tourists. The ethos of sustainability and integrity that has been stressed from the beginning has filtered through to other tourism entrepreneurs in the area. In that way, the Museum has to some extent catalysed a systematic approach to sustainable tourism development (Haven-Tang and Jones 2012) through which the small tourism companies in Strandir have been able to identify with. The commitment to sustainability is echoed by many local entrepreneurs who stressed in interviews that things needed to be properly done in order to provide what they labelled as an authentic experience. This is also reflected in the ideas and vision for future development of tourism in the area. Thus, there is a strong resistance to any ideas of mega developments that would not fit what people frequently labelled 'the character of Strandir' or of the environment where things tend to happen in a rather slow pace. On the other hand, some felt that a serious investment in tourism was needed in Strandir and a more focused tourism strategy should be on the agenda of the municipal government. The need of a 'real' hotel in Hólmavík was frequently mentioned along with the need for more recreational varieties. Tourism development in Strandir thus also invokes controversies and different accounts that may be challenging to weave together in the future.

Continuing the Story

The establishment of the Museum of Icelandic Sorcery and Witchcraft has actively changed the landscape of tourism in Strandir and played a central role in a turn towards tourism in the region. It is an effort to put tourism into use as a tool for economic diversification and cultural enhancement. This chapter has emphasized the entrepreneurial aspect of the story of the Museum, its ramifications and how it relates to different narratives of the region and of tourism. We have illustrated that the tactics underlying the establishment of the Museum were partly based on coincidence and dependant on improvised relational ordering of diverse entities such as narratives of witchcraft tightly tied to natural landscapes, policy discourses and personal engagement. The same can be said about many of the tourism entrepreneurs in the region that ventured into the sector almost by chance. While this is a characteristic of lifestyle entrepreneurship, the accounts we have followed illustrate how collective and institutional aspects are also of significance.

Many see the further development of low key and personal service as a favourable path to follow. While the personal contact and local insight provided by the entrepreneurs is a highly attractive 'product', it places particular demands on the providers. Hence, without further research and policy amendments that take the creative and improvisational character of rural tourism seriously, the long-term viability of ventures based on personal contact and co-created experiences cannot be confirmed.

As an example of relational ordering the Museum is still a precarious assemblage. Its base is uncertain because it is highly dependent on the personal engagement of its director. Nevertheless, the Museum has paved the way for others to start up tourism firms in the area on the basis of creative thinking and knowledge enhanced by the collective entrepreneurship of Jón, Siggi and Magnús that also received institutional backing from both regional and national bodies. Tourist entrepreneurs in the region look towards the Museum and Strandagaldur as role models for successful mastering of tourism mobilities, the professional working methods and emphasis on sustainability. In this, the leadership of Strandagaldur is likely to have limited the negative effects of the alleged weaknesses of lifestyle entrepreneurship. Thus, we conclude that the Museum has created a venue for collective entrepreneurship in tourism that has enhanced the tourism development in the region as well as boosted the tourism providers' self-esteem without restricting tourism entrepreneurs to a particular framework or agenda. Hence, entrepreneurs have been able to capitalize on the network(ing) and experience of the Museum's organizers while finding their own ways about in evolving tourism services and products. Thus, although the theme of witchcraft and sorcery is very much related to Strandir it is only one thread in the continuing weaving of the region's narrative as a tourism destination.

Acknowledgment

First, we would like to thank the editors for inviting us to submit a chapter to this volume. Also we express our gratitude to all the tourism operators in Strandir that took part in the research and assisted us in various ways. In particular, we thank Sigurður Atlason, the director of the Museum of Icelandic Sorcery and Witchcraft for invaluable assistance throughout the project. Finally we would like to thank our friend and research partner, Katrín Anna Lund for valuable comments on an earlier draft.

References

Ateljevic, J. (2009), 'Tourism Entrepreneurship and Regional Development', in Ateljevic and Page (eds).

Ateljevic, J. and Li, L. (2009), 'Tourism Entrepreneurship – Concepts and Issues' in Ateljevic and Page (eds).

Ateljevic, J. and Page, S.J. (eds) (2009), *Tourism and Entrepreneurship: International Perspectives* (Oxford: Butterworth-Heinemann).

Bauman, Z. (2001), *The Individualized Society* (Cambridge: Polity Press).

Benediktsson, K. (2001), 'Beyond Productivism: Regulatory Changes and Their Outcomes in Icelandic Farming', in Kim et al. (eds).

Bosworth, G. and Farrell, H. (2011), 'Tourism Entrepreneurs in Northumberland', *Annals of Tourism Research* 38:4, 1474–1494.

Byggðastofnun (1999), *Byggðir á Íslandi: Aðgerðir í byggðamálum* (Byggðastofnun).

Bærenholdt, J.O. (2011), *Coping with Distances: Producing Nordic Atlantic Societies* (Oxford and New York: Berghahn Books).

Bærenholdt, J.O. and Granås, B. (eds) (2008), *Mobility and Place: Enacting Northern European Peripheries* (Aldershot: Ashgate).

Callon, M. (2004) 'The Role of Hybrid Communities and Socio-Technical Arrangements in the Participatory Design', *Journal of the Center for Information Studies* 5:3, 3–10.

Cawley, M., Gaffey, S. and Gilmore, D.A. (2002), 'Localization and Global Reach in Rural Tourism: Irish Evidence', *Tourist Studies* 2:1, 63–86.

de Laet, M. and Mol, A. (2000), 'The Zimbabwe Bush Pump: Mechanics of a Fluid Technology', *Social Studies of Science* 30:2, 225–263.

Edwardsson, R. and Rafnsson, M. (2006), 'Basque Whaling around Iceland. Archeological Investigation in Strákatangi, Steingrímsfjörður', *Strandagaldur: Museum of Icelandic Sorcery and Witchcraft* [website] <http://www. galdrasyning.is/baskarnir.pdf > accessed 31 October 2012.

Florida, R. (2004), *The Rise of the Creative Class: And How It's Transforming Work, Leisure, Community and Everyday Life* (New York: Basic Books).

Georg, E.W., Mair, H. and Reid, D.G. (2009), *Rural Tourism Development Localism and Cultural Change* (Bristol: Channel View Publications).

Getz, D. and Carlsen, J. (2000), 'Characteristics and Goals of Family and Owner-Operated Businesses in the Rural Tourism and Hospitality Sectors', *Tourism Management* 21:6, 547–560.

Graves-Brown, P. (ed.) (2000), *Matter, Materiality and Modern Culture* (London and New York: Routledge).

Gren, M. and Huijbens, E. (2012), 'Earth to Tourism!', *Annals of Tourism Research* 39:1, 155–170.

Gunnarsdóttir, G.Þ. (2011), 'Reflecting Images: The Front Page of Icelandic Tourism Brochures', in Ísleifsson and Chartier (eds).

Hall, D. et al. (eds) (2005), *Rural Tourism and Sustainable Businesses* (Clevedon: Channel View Publications).

Hallam, E. and Ingold, T. (eds) (2007), *Creativity and Cultural Improvisation* (Oxford: Berg).

Haven-Tang, C. and Jones, E. (2012), 'Local Leadership for Rural Tourism Development: A Case Study of Adventa, Monmouthshire, UK', *Tourism Management Perspectives* 4, 28–35.

Huijbens, E., Hjalager, A.-M., Björk, P., Nordin, S., and Flagstad, A. (2009), 'Sustaining Creative Entrepreneurship: The Role of Innovation Systems', in Ateljevic and Page (eds).

Ingold, T. (2000), 'Making Culture and Weaving the World', in Graves-Brown (ed.).

Ingold, T. and Hallam, E. (2007), 'Creativity and Cultural Improvisation. An Introduction', in Hallam and Ingold (eds).

Ioannides, D. and Petersen, T. (2003), 'Tourism 'Non-Entrepreneurship' in Peripheral Destinations: A Case Study of Small and Medium Tourism Enterprises on Bornholm, Denmark', *Tourism Geographies* 5:4, 408–435.

Ísleifsson, S.R. and Chartier, D. (eds) (2011), *Iceland and Images of the North* (Québec: Presses de l'Université du Québec and The Reykjavík Academy).

Jamal, T. and Robinson, M. (eds) (2009), *The SAGE Handbook of Tourism Studies* (Los Angeles: Sage).

Jóhannesson, G.T. (2012), '"To Get Things Done": A Relational Approach to Entrepreneurship', *Scandinavian Journal of Hospitality and Tourism* 12:2, 1–16.

Jóhannesson, G.T. and Huijbens, E.H. (2013), 'Tourism Resolving Crisis? Exploring Tourism Development in Iceland in the Wake of Economic Recession', in Müller et al. (eds).

Jónsson, J. (1996), *Ferðaþjónusta og þjóðmenning* (Hólmavík: Héraðsnefnd Strandasýslu).

Karlsdóttir, A. (2008), 'Not Sure About the Shore! Transformation Effects of Individual Transferable Quotas on Iceland›s Fishing Economy and Communities', *American Fisheries Society Symposium* 68, 99–117.

Karlsson, Á. (2011), *Menningartengdferðaþjónusta á Vestfjörðum*. M.A. University of Iceland. <http://hdl.handle.net/1946/7513>, accessed 15 November 2012.

Kim, K-H. et al. (eds.) (2001), *Developing Sustainable Rural Systems* (Pusan: PNU Press).

Lane, B. (2009), 'Rural Tourism: An Overview', in Jamal and Robinson (eds).

Lash, S. and Urry, J. (1994), *Economies of Signs and Space* (London, Thousand Oaks and New Delhi: Sage).

Lund, K.A. and Jóhannesson, G.T. (under review), 'Earthly Substances and Narrative Encounters: Poetics of Making a Tourism Destination', *Cultural Geographies*.

McLean, S. (2009), 'Stories and Cosmogonies: Imagining Creativity Beyond "Nature" and "Culture"', *Cultural Anthropology* 24:2, 213–245.

Ministry of Industry and Commerce (2005), *Vaxtarsamningur Vestfjarða til aukinnar samkeppnishæfni og sóknar, tillögur verkefnisstjórnar um byggðaáætlun fyrir Vestfirði* (Reykjavík: Iðnaðar- og viðskiptaráðuneytið).

Mitchell, M. and Hall, D. (2005), 'Rural Tourism as Sustainable Businesses: Key Themes and Issues' in Hall et al. (eds).

Müller, D. et al. (eds). (2013), *New Issues in Polar Tourism: Communities, Environments, Politics* (New York: Springer).

Nyseth, T. and Viken, A. (eds) (2009), *Place Reinvention: Northern Perspectives* (Farnham: Ashgate).

OECD (Organisation for Economic Cooperation and Development). (1994), *Tourism Strategies and Rural Development* <http://www.oecd.org/dataoecd/31/27/2755218.pdf>, accessed 2 October 2012.

Peters, M., Frehse, J. and Buhalis, D. (2009), 'The Importance of Lifestyle Entrepreneurship: A Conceptual Study of the Tourism Industry', *PASOS. Revista de Turismo y Patrimonio Cultural* 7:2, 393–405.

Rafnsson, M. (2003), *Angurapi: Um galdramál á Íslandi* (Hólmavík: Strandagaldur).

Richards, G. (2011), 'Creativity and Tourism: The State of the Art', *Annals of Tourism Research* 38:4, 1225–1253.

Richards, G. and Wilson, J. (2006), 'Developing Creativity in Tourist Experiences: A Solution to the Serial Reproduction of Culture?', *Tourism Management* 27, 1209–1223.

Roberts, L. and Hall, D. (2004), 'Consuming the Countryside: Marketing for "rural Tourism"', *Journal of Vacation Marketing* 10:3, 253–263.

Samgönguráðuneytið (2001), *Menningartengd ferðaþjónusta* (Reykjavík: Samgönguráðuneytið).

Shaw, G. and Williams, A.M. (2002), *Critical Issues in Tourism: A Geographical Perspective* (Oxford: Blackwell).

Sigurðsson, B. et al. (2007), *Silfur hafsins – Gull Íslands: Síldarsaga Íslendinga* (Vol. 2.) (Reykjavík: Nesútgáfan).

Statistics Iceland (2011), *Tourism Satellite Accounts 2009–2011* (Reykjavík: Hagstofa Íslands).

Statistics Iceland (n.d.), *Population/Municipalities.* <http://www.statice.is/Statistics/Population>, accessed March 10 2013.

Statistics Iceland, Hildur.Kristjansdottir@hagstofa.is, 2013. Overnights in Strandir. [email] Message to G. T. Jóhannesson (gtj@hi.is). (Personal communication 2 May 2013, 16:18).

Strobl, A. and Peters, M. (2013), 'Entrepreneurial Reputation in Destination Networks', *Annals of Tourism Research* 40, 59–82.

Tulinius, T. (ed.) (2008), *Galdramenn: Galdrar og samfélag á miðöldum* (Reykjavík: Hugvísindastofnun Háskóla Íslands).

Þorgrímsdóttir, S. (ed.) (2012), *Samfélag, atvinnulíf og* íbúaþróun í *byggðarlögum með viðvarandi fólksfækkun* (Sauðárkrókur: Byggðastofnun) [pdf] <http:// www.byggdastofnun.is/static/files/Skyrslur/Samfelag/Samfelag_atvinnulif_ og_ibuathroun_skyrslan_i_heild.pdf>, accessed 13 October 2012.

Chapter 7
Ski Resort Development.
Scripts and Phronesis

Arvid Viken

Introduction

Even though ski resort development is discussed in tourism research (Gill 2000, 211; Tuppen 2000; Flagestad and Hope 2001; Nordin and Svensson 2007), it is a topic that is still sparsely covered. In one of the more quoted articles in the field, Flagestad and Hope (2001) write about 'ski destination', not denoting resorts. And in the literature dealing with tourist resorts, the terms are frequently combined into 'resort destination' or 'destination resort'. Although there is a variety of definitions and applications (Brey et al. 2007), a resort often refers to a more limited geographical location than a destination, an enclave where people undertake preferred activities and are fully accommodated (Lew 2004). It is a place to which 'people go for vacations or recreation, and may be qualified with such adjectives as health, inland, seaside, mountain, or ski' (Wall 1995). In a review article, Brey et al. (2007, 415) present a functional classification of 'destination resort research' (ibid., 420). One is called 'industry analysis', the approach that is applied here. The focus in the chapter is on knowledge and models that have been applied in a particular ski resort development: knowledge that has the form of scripts known from literature or the field, and knowledge that can be characterized better as practical experience. This broad view of knowledge, with a continuum from formal academic theories to practical knowledge, dates back to Aristotle. One of the types of knowledge he identified he labelled *phronesis*, often defined as practical knowledge (cf. Flyvbjerg et al. 2012), and a type of knowledge that is said to be vital in dealing with tourism (Tribe 2002). This chapter focuses on the knowledge aspect of ski resort development, while understanding knowledge as scripts that practitioners have learned from formal training, or a practical field, and in addition their practical experience, denoted as phronesis.

The tourism development analysed in this chapter takes place in northern Norway, in Målselv, where a ski resort has been developed since 2006. Målselv is a neighbouring municipality of Tromsø, the biggest town in northern Norway. The development process is very much an entrepreneurial undertaking. The project manager had a background from real estate development and marketing, but also other entrepreneurs were involved. None of them were trained particularly in ski resort development, or in tourism. The focus here is on concepts, scripts and experiences applied in the short-term start-up process. It took less than two years between the essential decisions being made in autumn 2006, and the opening of the ski resort in January 2008. The chapter starts by presenting the theoretical perspectives; what is a script, and the Aristotelian typology of knowledge. This is followed by a description of the methods applied. The presentation from our study is organized by first presenting the start-up period, followed by identifying applied development models; resort as a growth factor, a corporate resort model, copycatting, a cooperative land-owner model, regulation performances, involvement of stakeholders, and a section presenting knowledge and models scarcely appearing in the process. Towards the end, the limitation of phronesis, and its tendency to be transformed into epistemic knowledge, is discussed.

Entrepreneurial Knowledge: Scripts and Phronesis

The question raised here is what kind of knowledge, in terms of scripts that have been in action in the development of a ski destination, was introduced by the entrepreneurs involved. There is always some prior knowledge to entrepreneurship, and most people perform within a 'knowledge corridor', that enables 'him/her to recognize certain opportunities, but not others (Venkataraman 1997)' (Shane 2000, 452). Thus, the background and experiences of entrepreneurs are important for how they see the case. A ski resort developer needs to judge if the idea is good, create a project out of the idea, market the idea, get investors involved, and bring the project from a start-up and project phase to an operative stage. This is more or less the same in all start-up processes (Harrison and Leitch 2005, 356). Part of the management of such a process is to bring in the best of suitable models; ski resorts have been established all over the world, models of how to construct lifts, cabins, hotels, roads and other infrastructure exist, ready to be adjusted and applied. Such models can also be called scripts. 'Scripts are essentially recipes, borrowed, followed, and modified by individuals to get things socially and materially done ... [s]cripts are meaningful and competent if they allow the user to act quickly within particular business and social settings, to achieve legitimate, competent, and powerful results' (Chiasson and Saunders 2005, 751). Basically there are two types of scripts, those that are described in literature, and those that circulate in the industrial field in question, but often they overlap. Many of the challenges in an entrepreneurial process do not have solutions that are scripted at all. And even if the tasks are scripted, there are always aspects that are not. Therefore, in most entrepreneurial processes, besides the scripts that can be adopted and adapted, there is a practical experience at work.

Aristotle made a distinction between three types of knowledge, calling them *episteme, techne* and *phronesis.* 'Whereas episteme resembles our ideal modern scientific project, *techne* and *phronesis* denote two alternative roles of intellectual work' (Flyvbjerg 2004, 286). *Techne* refers to (technical) skills, whereas *phronesis* is practical, action oriented, context-adapted, and moral judgements; it is a 'practical value-rationality' (ibid., 287). According to Nonanka and Toyana (2007, 378) '*Phronesis* is an intellectual virtue. Roughly translated today as prudence, ethics, practical wisdom or practical rationality, phronesis is generally understood as the ability to determine and undertake the best action in a specific situation to serve the common good'. Thus, there are ethics and normative elements involved. Scholars such as Tribe (2002), Jamal (2004) and Dredge et al. (2012) see *phronesis* as an important part of the knowledge base of tourism, emphasizing that it is a practical field. Tribe (2002) calls the modern tourism practitioner trained in the field of tourism a philosophical practitioner, as he is trained both in liberal arts, and in vocational skills, both involving reflection and action.

Phronesis is often about how science and skills are used or put together. Therefore it is a good description of entrepreneurship. For entrepreneurs the job is to put bits and pieces from different disciplines and practice fields together, to communicate with different experts, and to apply these types of knowledge in a practical setting. In developing a ski resort there are lots of practical operations that are more or less prescribed by knowledge in disciplines such as architecture, engineering, law, business administration, or tourism theory. This is the episteme type of knowledge, providing scripts on how to go about the doing. The 'engineering' and the craftsmanship are 'know how' needed for implementation, most often taken care of by different types of experts (cf. Bruner 1966). Expertise, in general but perhaps in particular in complex projects like a ski resort development, is close to phronesis. Far from all is scripted and much is left to the judgements of those in charge, based on experience and a practical logic – what is here called *phronesis.* Nonanka and Toyama (2007, 379), who write within the field of strategic management, claim that '[p]hronetic leaders use their sense of the details to 'see' or 'feel' the problems of their organizations as solvable within local constraints', and they 'must be able to synthesize contextual knowledge accumulated through experience, with universal knowledge gained through training'. They also see phronesis as a strategic management tool for knowledge-centred companies, and claim that it consists of six abilities:

> (1) the ability to make a judgement on 'goodness', (2) the ability to share contexts with others to create a shared space of knowledge we call *ba*, (3) the ability to grasp the essence of particulars situations/things, (4) the ability to reconstruct the particulars into universals and vice-versa using language/concepts/narratives, (5) the ability to use any necessary political means to realize concepts for the common good, and (6) the ability to foster *phronesis* in others to build a resilient organization. (Nonanka and Toyama 2007, 379)

As will be shown in the following, all these abilities have been applied in the case in question.

One way of looking at phronesis is to perceive it as a kind of residual knowledge, i.e. knowledge that is not episteme or skill, but that can be identified. Flyvbjerg (2004) claims that social sciences are more apt for understanding social processes and practice than more exact disciplines like natural sciences or engineering. Building on Flyvbjerg, Schram (2004, 423) also claims that 'the social sciences can distinctively produce the kind of knowledge that grows out of intimate familiarity with practice in contextual settings'. With an explorative or open-ended approach, the significance and character of phronesis can be revealed. However, and as will be discussed through the analysis of Målselv to follow, when this is done, it is a question whether it still is phronetic knowledge (Laitin 2003; Schram 2004). The point of departure here is that a script is a recipe for how to go about, taken from an established theory or practical field, whereas phronesis is a broader practical platform also involving personal experience, and normative and ethical aspects (see Meyers 2004).

Methodology

The study behind this chapter was about tourism destination development, and not explicitly about the knowledge base. This aspect was something that became interesting through our visits in the field.[1] It became very apparent that there were some theoretical models applied, but also that much was founded on work experiences related to the way such resorts had been developed and managed in other places. Thus, to catch the variety of scripts, models and solutions, we had to search for development approaches that included more than those we find in the books. We decided to add a grounded theory inspired working method into the study of these processes.

In acquiring an understanding of the dynamics within the resort development in Målselv, a picture of a variety of stakeholders emerged. Stakeholders can be understood both as shareholders, holders of stakes or interests, holders of knowledge and holders of status (Schmitter 2002), or according to Freeman (2004, 229), 'any group or individual that can affect or is affected by the achievement of a corporation's purpose'. Among the obvious stakeholders in this case are the ski resort development company, the municipality, the construction and supply businesses, the reindeer herders and the farmers, the army, the sports, culture and arts actors, the airport, but also citizens from different villages. These stakeholders have all been approached, and have been interviewed individually or in focus groups.

In the study behind this paper, the stakeholder perspective is also brought into the methodology, as stakeholder segmented focus group interviews have been the major research method (cf. Kontogianni et al. 2001). The informants were sorted in groups characterized by a particular attribute (like occupation, gender or age), supposed to influence their perspectives (Viken 2013). This is an explorative method suitable for creating an overview of topics and attitudes in

---

1   The research team consisted of Torill Nyseth, Anniken Førde, Per Kåre Jacobsen, and Arvid Viken.

Table 7.1   Stakeholders and their positions and interests

Stakeholders	Positions
Focus group 1: High School teachers	The public service sector
Focus group 2: Journalist, teachers, cultural workers	Several private and public institutions
Focus group 3: Arts and culture	Cultural industries
Focus group 4: Business actors	Private companies
Two Managers in Construction Company	Construction business
Ski Resort Entrepreneur together with an employee	Målselv Fjellandsby
Two welfare officers in the military	The military administration
Mayor	Municipality
Public planner	Municipality
Public business developer	Municipality
Business representative	Chamber of commerce
Airport manager	Airport
Destination company representative	Tourism industry
Sport and tourism actor	Sports hotel
Sport actor	Ski sport organizations
Statskog SF (state owned land)	Farmers
Reindeer herder	The reindeer herders district

an on-going development process, to understand projects and entrepreneurs, but is also a method to unveil how people have been thinking (Morgan 1988). Some stakeholders were more central to the process than others – such as the project manager and the mayor – and these were approached separately with in-depth interviews. The interviewees are as follows (Table 7.1).

All interviews lasted between one and a half to two hours, except for one that took four to five hours involving a dinner. Most of the interviews were transcribed afterwards (see also Førde in Chapter 9 of this volume).

The focus group interviews took the form of discussions on the topic given. For some, but not all of the interviewees, the topics discussed were part of their daily agenda. They were all stakeholders, but not equally involved or informed. There also were issues that were raised by us, that were not hitherto on the local agenda – for instance, concerning perceptions of place, branding and resort development perspectives. Thus, by using this method, there is a chance that the group sessions will reveal or start local debates. In this respect there is a chance that the researchers become players in the ongoing political processes, adding new topics or perspectives to the local agenda (see Viken 2013). The value – positively or negatively – of this, is not known.

The Start-up Story: Entrepreneurship and Private-Public Partnership

During the Cold War period Målselv was a central military base with thousands of soldiers and more than 8500 inhabitants. With the fall of the Soviet Union, Målselv lost its strategic significance, and the Norwegian Ministry of Defence decided to down-scale its activities in the area. As this happened, the number of inhabitants decreased, and so did the municipal tax revenue. Thus, around the turn of the millennium, people were nervous about the future. The municipality of Målselv took the lead in the search for alternative industries. Among other things a yearly event called the Inland Conference had discussed different industrial opportunities, including tourism. Involvement from the authorities in such processes is known from many other places, for instance, Whistler in Canada. Gill (2000) has shown how the municipal authorities were central in the start-up period of the ski resort, for example by creating a land-use plan. Also the development story of Åre in Sweden shows the importance of the public sector in policymaking and in forming strategies for ski tourism development (Nordin and Svensson 2007). In recent years an external private corporation, Skistar, has been the dominant actor in Åre, in charge of ski-lifts, booking and parts of the accommodation, professionalizing and commercializing the ski tourism, it is argued (cf. von Sydow 2007; Nordin and Westlund 2009). In Målselv the municipality still in 2013 is supposed to have a vital role, due to recession and economic problems, in the post start-up period.[2]

Informants in Målselv identify one event of particular significance for the start-up of the ski resort development. One of the municipal officers happened to know of a tourism entrepreneur and resort developer from Voss, a ski destination in the south of Norway. He was invited to Målselv, primarily to look at some hotel projects related to the river, and ski resort opportunities. According to informants, he was impressed with what he saw, and particularly how well suited the area would be for downhill skiing. He decided to become involved, and a fast-going process started.

To develop the ski resort, the entrepreneur (later called the project manager), established a company called Målselv Development. The company had a rather small starting capital of NOK 100 000. After some months an issue was made, and the capital base increased to NOK 150 000. The newcomers had to pay NOK 12 million for half of the company's shares. Six local investors, that some months earlier had created the investment company Målselv Invest, put NOK eight million

---

2   In 2013, Målselv Development went bankrupt, due to lack of sales of lots and cabins; only about 40 had been sold in the last three years.

on the table, and two investors from Tromsø filled up the rest of the issue. In reality the issue provided Målselv Development with a financial platform valued at NOK 24 million. The involvement of local investors was an important signal to the community, but it was also an investment that could give them pay-back; they were all in the construction sector. On the financial side, another important act was a loan of NOK 60 million taken by the municipality, for development of the infrastructure, including a road, cabling, and the ski lifts. The agreement was that for each lot sold some of the money should go to redemption of the loan. Cabins were sold on prospects, which was also part of the financing of the project. Although affected by a recession in 2009, by summer 2011 around 250 lots were sold, 148 cabins built, and 48 flats had been constructed. The future ambitions were to sell 1500 lots. Altogether about NOK 600 million was invested during the five first years. However, the efforts to catch the interest of a major hotel investor have not paid off so far at the time of writing in 2013, although a hotel is strongly needed for the ski market. To attract people to buy lots or ready-made cabins, a number of ski slopes and three lifts were constructed – in sum, a small ski resort.

The mayor at the time (2003–2011) played an important role in the Målselv case. He can be seen as an entrepreneur also. This is a phenomenon is known from the literature; mayors and senior local government officials operating as politician-public entrepreneurs (Zerbinati and Souitaris 2005). They use their entrepreneurial skills to build a network, and to gather political support in order to acquire resources and funding for local projects. This is also a good description of the mayor in Målselv. He was eager to find alternative industries and jobs for the community when the military announced its partial withdrawal. He followed a series of paths, one of which was about tourism. The mayor's strengths were political skills and networking. Concerning the ski resort development, he enabled a fast decision process within the municipality, and he involved the county councilor particularly responsible for environmental issues and made him an ally in the process. The mayor also had contacts in the military and the Labour party, and had access to the Minister of Defense. He definitely acted as a facilitator in the process (cf. Johannisson 1990), and together with the project manager he pushed the development process.

To start a resort development, there is a need for some early or pioneer customers. Certain types of people were approached in the starting period – among these well situated and high profile people from Tromsø (two hours' drive away). To recruit these people, the prices of lots were kept low. This is a known strategy in tourism (cf. Lee et al. 2008); to get some prominent persons, known from business, sports, arts or media, to front a project or new venture. Part of this strategy is also to identify the so-called 'critical mass'; the type and amount of customers that are needed for break even. According to Oliver et al. (1985, 547) '[a] small core of interested and resourceful people can begin contributions toward an action that will tend to 'explode', to draw in the other, less interested or less resourceful members of the population and to carry the event toward its maximum potential'. Thus, to develop a ski resort is very much about taking the right decisions, provoking essential actions, and involving the right people.

The start-up in Målselv followed well known scripts; it is about entrepreneurship, financing models, public-private partnership, creating a customer base and a critical mass. However, the critical mass was not established – the local and regional market seems to be too small, and an international recession with national impact occurred during the start-up period. No script seems really to prescribe how such situations should be handled.

Scripts and Phronesis Performing in Målselv Fjellandsby

Resort development as growth

The project manager in Målselv sees himself as a real estate developer. His view was that the profit lies in developing lots, flats and cabins and other accommodation, not in ski lifts and infrastructure. This has also been recognized in other places; for most investors and developers the real estate part is the core, and ski services are of secondary importance (Clydesdale 2007). The municipality of Målselv had a focus on growth, or rather, activities that could compensate for the reduction in military activities. The growth perspective is well known in destination and resort development (Viken in Chapter 2 of this volume). Particularly the term 'growth machine' has been used in studies of destination development (Canan and Hennesey 1989; Gill 2000). One of these studies focuses on the development of Whistler, Canada, in the 1970s and 1980s. According to Gill, a growth machine occurs when:

> those who stand to gain directly as landowners, speculators, or investors form local growth coalitions with service providers (such as realtors, bankers, and local media) to promote overall growth (Canan and Hennesy 1989). These local growth elites play a major role in supporting local politicians, who also view growth as a key function of local government. (2000, 1083)

This is in fact a summary applicable to the case study presented here. Local investors, speculators and authorities made alliances with an entrepreneur and local authorities to start a ski resort development to secure local growth. The reason why a ski resort was chosen, and in particular the real estate part of it, was the profit and growth potential. The project manager's focus was to make profit through developing and trading lots, flats and cabins. When such a development is started, the prices and values of real estate tend to increase. In developing a tourist resort, the land changes character from being valued for its production potential of crops, timber or grazing land, to being assessed by its exchange value in the leisure and tourism markets. Overvåg (2010, 7) shows that after a second-home resort development in a rural area in Norway (Ringebu), the prices had more than tripled during an eight-year period. In Målselv interviewees mentioned that it has become many times as expensive to build a cabin in this area. Second-home development normally also means opportunities to make money by selling land, and work for the construction industry (Rye 2011). In Målselv this industry is significant, primarily due to continuous military construction activities. In addition, tourism development brings external money into the local economy (Bieger et al. 2007; Rye 2011), by investors in the project, second-home buyers, and tourists. There are of course also some contested impacts of such a tourism development (Tuulentie and Mettiäinnen 2007; Rye 2011), among them, that prices tend to rise, and changing lifestyles (Rye 2011, 3). Thus, over the years, growth management tends to become a crucial issue (Gill and Williams 2011). The growth paradigm and its paradoxes are well known in the literature in the fields of social economics, political geography and sociology, and the ideas are also part of a public political discourse on regional development, and a goal for places in transition. There are many models, places that have succeeded in creating positive economic trends by establishing ski resorts. Whistler in Canada and Åre in Sweden have been mentioned, others are Levi, Ylles and Sarisälkä in Finland, Trysil and Hemsedal in Norway.

The ski resort: A corporate or a community destination model?

Målselv Fjellandsby is located up in the mountains, 14 kilometres from the community centre. The area was a forest and grazing land with nothing else. As a resort it has been transformed to become a social and economic system that supports its customers by supplying most of their needs. The development of the ski resort as described above has primarily been managed by the development company that was established for this purpose. This fits with what is known as the corporate destination model (Flagestad and Hope 2001). The alternative is to base a resort within an existing community, more or less as an integrated part of local everyday life. In the corporate development model, the tourism site and production are organized as a firm that is in charge of the slopes, the ski-lifts, ski outlet, ski school, booking, some of the accommodation, and so on. This is also called integrated resort management, and has proved to be more efficient from a technical or economic point of view (Clydesdale 2007; Falk 2009). As the operation in Målselv has a small scale, the resort developers have welcomed other actors into the area – they have invited several hotel chains to build a hotel within the resort. The theoretical stance concerning the corporate versus the community approaches is basically found in tourism theory, but one of its conveyors is Norwegian, and the script is also well known and discussed in the ski resort field, known as the SkiStar model. SkiStar is a resort company in charge of major ski resorts both in Norway and Sweden run on the corporate destination model, discussed here.

The business concept: Copy-catting Levi in Finland

Imitation is a well-known entrepreneurial pattern. In fact to apply known business concepts or development models has the flavour of copying. In Målselv the developer has in particular tried to copy Finland, the municipality shares a long border with this country. Winter tourism development in Finland has been a success since the 1980s. There are a whole series of ski resorts, including Levi, Ylläs, Saarisälkä and Ruka, all of a significant size. No such resorts exist in northern Norway.

Our study has confirmed that thoughts about why the Finnish ski resorts have succeeded, a certain envy of such success stories, and ideas of copying them, have been present in and influenced on the development of Målselv Fjellandsby. The prerequisites for ski resort development are very different in the two countries. In Norway the best conditions for ski resort development have been in the south, near the most populated areas (ski resorts at Norefjell, Geilo, Hemsedal, Hafjell, Trysil, Oppdal, Voss). In Finland the situation is different, with no mountains near the big cities in the south. All the major Finnish ski resorts are in the north, where domestic tourism has paved the way for a certain foreign market success. The model for Målselv was primarily Levi, the most successful resort in Finland, and several study tours across the border have been made. Both the skiing conditions and the resource base should be as good in Målselv as in Levi, the developers maintain. They openly have talked about the copycatting in interviews, speeches and media. 'If Finland can, we can', is their claim. It is a transparent copycatting.

The idea was to copy everything; the resort, the branding, the target market profile. Most interesting is the copying of the brand (Førde in Chapter 9 of this volume). Santa Claus has been a branding figure of northern Finland for years. The developers in Målselv found a similar figure; the Snowman. They renamed the resort Snowman Resort and the marketing company as Snowman Booking, and even the airport, basically a military airport, has been given the name Snowman International Airport, inaugurated in March 2011, with a Norwegian minister and the county councillor attending the event. One year later this strategy was abandoned, and the resort renamed Målselv Fjellandsby. It was only the airport authorities that still promoted Målselv as the Snowman Resort two years afterwards.

Copycatting is to use existing ventures as scripts. It is a method known from theory, but also a practical undertaking. It is to learn from the practice of others. This is a phronetic learning process. There are also elements in the Finnish script that is not easy to copy for impatient developers; such as the time it has taken, 20–30 years or more. Projects like the one described here tend not to give room for time-consuming processes.

The land-owner cooperation: Applying the Trysil model

To develop a ski resort, the developer must get hold of the land needed. There are of course a whole series of alternative models for this, also depending on who the owners are, what the land has been used for hitherto, and so on. Raising the subject of tourism development may well be seen as a threat by the land owners. The model applied in Målselv has been used in several places – the best known is Trysil, the biggest ski resort in Norway, but it seems not to be formally described. The model is that the owners create a real estate company that governing an area with several owners, and hand the management over to a resort management company. In Målselv the owners were nine farmers, and the State Forest (a company managing state land). In addition, reindeer herders had rights to graze land in the area. Before the project manager approached the land owners, he had relatively clear ideas of how the resort ownership model should look like. He held a meeting where he showed a hand-made drawing of the area and presented his ideas including the model for ownership cooperation. The owners organized a real estate company. In Trysil the parties received a share proportional with the size of the land they owned within the area, but here, he insisted on all nine farmers having *equal* shares. In Målselv two more parties were added, the State Forest, and the reindeer herders with herds in the area. With this model the ground has the same value whether it is used for a lot, a road, a lift, a slope, or just an open area. Implicit in the model is that the land-owners still own the land, but have no influence on how it is used. This is taken care of by a development company (Målselv Development). The parties receive payment for the ground, after a formula they have agreed upon. During the first four years all of the landowners, including the reindeer herders, had received more than NOK one million each. However, the model is disputed; the landowners blame the company for not selling as many lots and cabins as expected, and the development company has asked for renegotiation of the agreement as the original deal is too expensive, as the company sees it.

The model described here is well known among people dealing with ski resort development in Norway. The ownership model can be seen as part of, or a prerequisite for, the corporate destination model, described above. To run a resort as a company, some sort of collective ownership has to be established. However, it can also be seen as a script known from the field, or part of a common phronesis of resort development people. Involved in the model is not only a technical description, but ethical and normative considerations, for instance, concerning who to involve, the (equal) size of the shares, and how to go about organizing.

To produce an area regulation plan: A consultant, lobbying and pressure

To develop a ski resort, there are many formal requirements regarding regulation that have to be dealt with, most of them prescribed in the Municipal and Regional Planning Act (*Plan og bygningsloven*). According to this law, the area should be planned, which means that it should be declared for what purposes different parts of the area would be used, and this must be based on an assessment of consequences for different stakeholders and the environment. The regulation plan was produced by an externally based company (called Opus), that the project manager had used in a ski resort project in which he had previously been involved. The consequence reports were made by a national environment assessment institute (Danielsen and Tømmervik 2006; Granmo and Tømmervik 2006; Jacobsen and Bjerke 2006). The regulations were done according to the requirements. As part of the process the reindeer herders' area board (*Områdestyret*) had to acclaim the proposals before the municipal assembly could endorse the regulation plan. The municipal assembly and the reindeer herding area board happened to have meetings the same day, and the municipal assembly had to wait for the area board's decision to endorse the overall regulation plan. At that time, those involved had gathered where the construction works should start, and the machines were lined up. When the decisions had been made, it was celebrated with champagne and the work started. The project was on track. It obviously was seen as an important event and project, locally.

The process described here is based on systemic expertise combined with local knowledge. The expertise can be identified as scholarly knowledge within planning and law, combined with insights about local circumstances. The local knowledge in action is, and in accordance with what has been observed by Yanow (2004), context-specific, interactively derived, lived-experience based, and partly tacit. Those involved knew what the minimum of assessments and regulations needed was, and also about possible obstacles in such a process. Their experience, for instance, told them to establish a tight relation to the county councillor responsible for the environmental issues. Yanow (2004) characterizes such knowledge as a sort of phronesis. In a way, the scripts in action seemingly had a phronetic wrapping.

To avoid conflicts: Managing stakeholders

According to the project manager, a major strategy in the process has been collaboration. This implies recognition of a whole set of stakeholders, seeing the resort development process as a stakeholder matter (cf. Perdue 2003). Two stakeholders were particularly important to tackle; the reindeer herders and the military. The losses of the reindeer herders were small, but likely to increase if the resort expanded. The gains were a cash compensation, and a right and space for a reindeer and Sami-based tourism operation within the resort village, and incomes as shareholders in Målselv Real Estate. To include the reindeer herders as a partner in this organization was symbolically important, recognizing their rights to the area. The project manager and the authorities did not make the mistake of neglecting Sami interests. The philosophy of the project manager seems to have been that business is business, whoever is behind it. However, this does not mean that all parties were equally involved (Førde in Chapter 9 of this volume).

With a central position in the Norwegian and NATO military strategies, the military presence had for decades been significant in Målselv. There might be a perception of military presence and tourism being a bad match, but this is not necessary the case. The Norwegian military has for some years had a philosophy of community involvement. Thus, when the decision about developing a ski resort was made, the military turned out to be an ally, and a stakeholder in the development of the tourist resort. There is a huge combined military-civilian airport in town. In working to establish the Snowman Resort, the airport was renamed Snowman International Airport, and given a make-over, with a big snowman painted on the front of the terminal building, obviously, acclaimed by the military authorities.

Also in the daily operation of the ski resort the military forces are important as a customer – the facilities are used for training and soldier welfare.

The need for collaboration and involvement seemed to have been acknowledged both by the project manager and the mayor. There is a strong theoretical support for this; collaboration, networks, network-based governance and adaptive management are among the buzz-words in contemporary public administration and its theoretical realm. It is part of the neoliberal governing paradigm. It probably is an example of how phronesis interacts with episteme, the practical wisdom is mixed with well-established theories.

Knowledge that Slipped Away in the Process

To develop a ski resort is also to create something physical; slopes, lifts, roads, lots and cabins, other facilities and infrastructure. As mentioned earlier, entrepreneurship is related to prior experience and to the knowledge system the actors are part of (Venkataraman 1997). In the case in focus, the technical sides of ski resorts were not a major priority. It seems to have been outside the project manager's 'knowledge corridor' (ibid.). This is said to be a limitation of most expertise (cf. Hinds and Pfeffer 2001). But scripts also exist in this area. The ignorance of these can in fact be observed. For instance, the ski lift splits the major slope into two narrow sections, instead of being on one side, and the cafeteria view faces the lift area instead of the beautiful valley below, along with many minor faults of more technical type. It was also openly admitted by the developer that no training or recruitment plan had been developed, and the competency in running a ski resort has not been prioritized. The project could obviously have been better anchored in the community. A ski resort close to a local community affects people's lives. They have been informed, but not involved. As demonstrated by Førde (Chapter 9 of this volume), particularly the place branding part concerns the local community. The developers blame their faults on the speed in the process, and themselves. As the project involved a significant amount of public funding and involvement, it could be questioned if it should be allowed to make such faults – it can be seen as waste of taxpayers' money. Such errors are surely related to the type of knowledge involved; the project manager is experienced in real estate development and marketing, not aware of, or not having the capacity to take into account, all the expertise and scripts that existed. The more technical aspects were neither part of his phronesis, nor a part of the competency of the construction companies. Thus, some of the faults and shortcuts made can probably be blamed on the phronetic approach identified. A consequence is a product and services not as good as they could have been. Phronetic knowledge has its confines.

Concluding Remarks: Knowledge on Track or Off-piste?

The focus in this chapter has been on knowledge applied in a ski resort development process.

There is a stock of knowledge, what can be seen as *epistemic* knowledge in the field, research-based knowledge about the real estate platform, the corporate model, the involvement of authorities, and about stakeholder involvement. Concerning the resort development studied here, there are also scripts applied originating from the field, concerning planning and how to organize the landowners. In the construction, but also in running a resort, there is a lot of *techne* or skill type of knowledge involved. However, how to put this together is not found in books or be acquired through formal training. Somehow, phronesis can be seen as compensation for lack of episteme and techne. For entrepreneurs this often seems to be the major type of knowledge, although it is difficult to measure how important the phronetic element is.

Theory, according to Robinson (1998), provides models and terms necessary in perceiving practice. But theory is not equal practice. This is also why universities rather seldom create ready-made practitioners, and internship is seen as a natural part of the study programme in many fields (Jamal 2004). The residual knowledge has to be acquired through practice. Therefore, one can never fully be taught how to make a ski resort. The task is too complex and context-dependent. The phronesis involved enables entrepreneurs to handle such situations, but the approach has its shortcomings. In the case described here, the project manager recognized the opportunities of the place, realized that a regional market existed, and convinced others that the idea was good. He applied models from other places, and knew how to handle the project politically. As Nonanka and Toyama (2007, 379) claim for entrepreneurs in general, the project manager was even able to 'foster phronesis in others to build a resilient organization', and seemed to have had most of the phronetic abilities that Nonanka and Toyama (ibid.) list as strategic management tools.

There are reasons to believe that phronetic knowledge has a more central place in the construction of modern knowledge today than earlier. To a greater extent than before knowledge seems to be produced through negotiations and collaboration between the knowledge producers and the industrial fields. This matches a contemporary view of knowledge and its production, called Mode 2 knowledge, described by Gibbons et al. (1994). Mode 2 knowledge is 'heterarchical, transient, transdisciplinary, socially accountable and reflexive, and undertaken in a context of application' (Goddard and Chatterton 1999, 687), differing from traditional disciplinary, hierarchical knowledge (Mode 1). The Mode 2 knowledge is said to be more innovative and adaptive, and is often developed in dialogue between the industries and academia. As a consequence, the borders between practical and academic knowledge are becoming more and more blurred. As this study has shown, many of the scripts and models are applied in developing the ski resort, known from the literature, but also among practitioners. The phronesis type of knowledge identified here circulates in a field where also knowledge producers and scholars take part. Maybe a new form of phronesis is emerging where the practical field and academy are integrated, and the boundaries between diminishing. However, there are reasons to believe that the comprehension varies. As this study has shown, the phronesis in action has had its limitations, stories of faults and fiascos frequently appear. Phronesis is about trying out, making errors, and learning what the books never tell. It can also be that the merger of episteme and phronesis primarily takes place on the surface. In these fields and within business and tourism in general, fads and fashions are performing (see e.g. Ogbonna and Harris 2002). Terms and theories are picked up and used, often without a particularly deep insight. May be it is not so much knowledge that merge, but the knowledge producers and potential users that move in the same circles.

One of the issues implicit in this discussion is what happens when phronetic knowledge becomes a matter of research, what type of knowledge is it then? Is identifying phronesis a way of transforming it into epistemic knowledge? Or will phronesis always be present, as a kind of practical complement to formal training and scripts? Probably, and as Meyer (2007) states, phronesis is needed 'to be able to analyze problems, to distinguish among more and less relevant facts, to know the best means for achieving desired ends, and, maybe most important, to know where to seek advice'. Or to put it another way: ethics, social and cultural norms are seldom scripted, but culturally inscribed. Thus, phronesis is not only a kind of knowledge, it is a moral advocate.

References

Bruner, J. (1966), *The Culture of Education* (Cambridge, MA: Harvard University Press).

Canan, P. and Hennessy, M. (1989), 'The Growth Machine, Tourism and the Selling of Culture', *Sociological Perspectives* 32:2, 227–243.

Chiasson, C. and Saunders, C. (2005), 'Reconciling Diverse Approaches to Opportunity Research Using the Structural Theory', *Journal of Business Venturing* 20:6, 747–767.

Clydesdale, G. (2007), 'Ski Development and Strategy', *Tourism and Hospitality Planning and Development* 4:1, 1–23.

Danielsen, E. and Tømmervik, H. (2006), Målselv Fjellandsby. *Konsekvensutredning, deltema reindrift*. NINA Report No. 179. (Tromsø: NINA).

Dredge, D., Benckendorff, P., Day, M., Gross, M.J., Walo, M., Weeks, P. and Whitelo, P. (2012), 'The Philosophic Practitioner and the Curriculum Space', *Annals of Tourism Research* 39:4, 2154–2176.

Falk, M. (2009), 'Are Multi-Resort Ski Conglomerates More Efficient?', *Managerial and Decision Economics* 30:8, 529–538.

Flagestad, A. and Hope, C.A. (2001), 'Strategic Success in Winter Sports Destinations: A Sustainable Value Creation Perspective', *Tourism Management* 22:5, 445–461.

Flyvbjerg, B. (2004), 'Phronetic Planning Research: Theoretical and Methodological Reflections', *Planning Theory and Practice* 5:5, 283–306.

Flyvbjerg, B., Landman, T. and Schram, S. (2012), *Real Social Science. Applied Phronesis* (Cambridge: Cambridge University Press).

Førde, A. et al. (eds) (2013), *Å finne sted. Metodologiske perspektiver i stedsanalyse* (Trondheim: Akademika forlag).

Freeman, R.E. (2006), *A Stakeholder Approach to Strategic Management,* Working paper No. 01–02. Darden Graduate School of Business Administration. (Charlottesville: University of Virginia).

Gibbons, M., Limoges, C., Nowotny, H., Trow, M., Scott, P. and Schwartzmand, S. (1994), *The New Production of Knowledge* (London: Sage).

Gill, A.M. (2000), 'From Growth Machine to Growth Management: The Dynamics of Resort Development in Whistler, British Columbia', *Environment and Planning A* 32:6, 1083–1103.

Gill, A.M. and Williams, P.W. (2011), 'Rethinking Resort Growth: Understanding Evolving Governance Strategies in Whistler, British Columbia', *Journal of Sustainable Tourism* 19:4/5, 629–648.

Goddard, J. B. and Chatterton, P. (1999), 'Regional Development Agencies and the Knowledge Economy: Harnessing the Potential of Universities', *Environment and Planning C: Government and Policy* 17:6, 685–699.

Granmo, L.P. and Tømmervik, H. (2006), *Konsekvensutredning, deltema landbruk.* NINA Report nr. 180 (Tromsø: NINA).

Grote, J.R. and Gbikpi, B. (eds) (2002), *Participatory Governance* (Opladen: Leske-Budrich).

Harrison, R.T. and Leitch, C.M. (2005), 'Entrepreneurial Learning: Researching the Interface between Learning and the Entrepreneurial Context', *Entrepreneurship Theory and Practice* 29:4, 351–371.

Hinds P.J. and Pfeffer, J. (2001), *Why Organisations Don't 'Know What They Know': Cognitive and Motivational Factors Affecting the Transfer of Expertise.* Research Paper No. 1697 (Stanford: Stanford University).

Jacobsen, K.O. and Bjerke, J.W. (2006), *Målselv Fjellandsby. Konsekvensutredning, deltema naturmiljø.* NINA Rport No. 176 (Tromsø: NINA).

Jamal, T.B. (2004), 'Virtue Ethics and Sustainable Tourism Pedagogy: Phronesis, Principles and Practice', *Journal of Sustainable Tourism,* 12:6, 530–545.

Johannisson, B. (1990), 'Community Entrepreneurship – Cases and Conceptualisation', *Entrepreneurship and Regional Development* 2, 71–78.

Katz, J. and Brockhaus, R. (eds) (1997), *Advances in Entrepreneurship, Firm Emergence and Growth* (Greenwich, CT: JAI Press).

Kontogianni, A., Skoutor, M.S., Langford, I.H., Bateman, I.J. and Georgiou, S. (2001), 'Integrating Stakeholder Analysis in Non-Market Valuation of Environmental Assets', *Ecological Economics* 37:1, 123–138.

Laitin, D.D. (2003), 'The Perestroikan Challenge to Social Science', *Politics and Society* 31:1, 163–183.

Lee, S., Scott, D. and Kim, H. (2008), 'Celebrity Fan Involvement in Destination Perceptions', *Annals of Tourism Research* 35:3, 809–832.

Lew, A.A. (2004), 'Editorial: Tourism Enclaves in Place and Mind', *Tourism Geographies* 6:1, 1–1.

Meyers, C. (2007), 'Institutional Culture and Individual Behavior: Creating an Ethical Environment', *Science and Engineering Ethics* 10:2, 269–276.

Morgan, D.L. (1988), *Focus Groups as Qualitative Research* (Thousand Oaks: Sage).

Nonaka, I. and Toyama, R. (2007), 'Strategic Management as Distributed Practical Wisdom (Phronesis)', *Industrial and Corporate Change* 16:3, 371–394.

Nordin, S. and Svensson, B. (2007), 'Innovative Destination Governance', *Entrepreneurship and Innovation* 8:1, 53–66.

Nordin, S. and Westlund, H. (2009), 'Social Capital and the the life Cycle Model: The Transformation of the Destination of Åre', *Tourism Review* 57:3, 259–284.

Ogbonna, E. and Harris, L.C. (2002), 'The Performance Implications of Management Fads and Fashions: An Empirical Study, *Journal of Strategic Marketing* 10:1, 47–68.

Oliver, P., Marwell, G. and Teixeira, R. (1985), 'A Theory of the Critical Mass. Interdependence, Group Heterogeneity, and the Production of Collective Action', *The American Journal of Sociology* 91:3, 522–556.

Overvåg, K. (2010), 'Second Homes and Maximum Yield in Marginal Land. The Re-resourcing of Rural Land in Norway', *European Urban and Regional Studies* 17:1, 3–16.

Perdue, R.R. (2003), 'Stakeholder Analysis in Colorado Ski Resort Communities', *Tourism Analysis* 8:2/4, 233–236.

Robinson, V.M.J. (1998), 'Methodology and the Research-Practice Gap', *Educational Researcher* 27:1, 17–26.

Rye, J.F. (2011), 'Conflicts and Contestations: Rural Populations' Perspectives on the Second Home Phenomenon', *Journal of Rural Studies* 27:3, 263–274.

Schmitter, P. (2002), 'Participation in Governance Arrangements: Is There a Reason to Expect It Will Achieve 'Sustainable and Innovative Policies in a Multi-Level Context?', in Grote and Gbikpi (eds).

Schram, S.F. (2004), 'Beyond Paradigm: Resisting the Assimilation of Phronetic Social Science', *Politics and Society* 32:3, 417–433.

Shane, S. (2000), 'Prior Knowledge and the Discovery of Entrepreneurial Opportunities', *Organization Science* 11:4, 448–469.

Tribe, J. (2002), 'The Philosophic Practitioner', *Annals of Tourism Research* 29:2, 338–357.

Tuulentie, S. and Mettiäinen, I. (2007), 'Local Participation in the Evolution of Ski Resorts: The Case of Ylläs and Levi in Finnish Lapland', *Forest, Snow, Landscape Research* 81:1/2, 207–222.

Tuppen, J. (2000), 'Restructuring Winter Sports Resorts', *International Journal of Tourism Research* 2:5, 327–344.

Venkatarman, S. (1997), The Distinctive Domain of Entrepreneurship Re-search: An Editor's Perspective, in Katz and Brockhaus (eds).

Viken, A. (2013), 'Fokus på sted-fokusgruppesamtaler som metode i steds- og destinasjonsanalyser', in Førde et al. (eds).

von Sydow, K. (2007), *Att slänga bensin där det redan brinner – En studie av expansionen i Åre by och turismens påverkan på Åres samhällsutveckling.* Arbeidsrapporter, Kulturgeografiska institutionen, No. 625 (Uppsala: Universitetet i Uppsala).

Wall, G. (1996), 'Integrating Integrated Resorts', *Annals of Tourism Research* 23:3, 713–717.

Yanow, D. (2004), 'Translating Local Knowledge at Organizational Peripheries', *British Journal of Management* 15:1, 9–25.

Zerbinati, S. and Souitaris, V. (2005), 'Entrepreneurship in the Public Sector: A Framework of Analysis in European Local Governments', *Entrepreneurship and Regional Development* 17:1, 43–64.

Chapter 8

Sled Dog Racing and Tourism Development in Finnmark. A Symbiotic Relationship

Kari Jæger and Arvid Viken

Introduction

This chapter explores the development of Europe's longest sled dog race, the Finnmark Race (*Finnmarksløpet*), and its significance for winter tourism in North Norway. As the name indicates, the race takes place in Finnmark, the northernmost county of Norway, where more than a hundred participants take between three and seven days to cross the Finnmark Plateau, depending on race category and performance. In recent years, the race has been broadcasted on the television and the internet, and has become well known by the general public in Norway. The race and Finnmark as a winter destination have developed as two intertwined trajectories, with many types of interrelations and reciprocal gains. While sporting events tend to have significant effects on tourist destinations (Gibson 1998; Zauhar 2004), this case tells of reciprocal effects, implying that tourism is also a platform for sporting events. Both the race and dog sled tourism have been important factors in the emergence of Finnmark as a winter tourism destination.

The practice of dogsledding is rather new in Norway, and in Finnmark it dates back to the 1970s. In 1925, Leonhard Seppala and Gunnar Kaasen, originating from the area, became American celebrities after having organized and participated in a sled dog relay that brought medicine to diseased gold-diggers in Nome, Alaska, all the way from Anchorage. This event prepared the ground for the dog sled race Iditarod, the longest and most popular dog sled race in the world, which runs between the same two towns. In Alaska, dogsledding is embedded in the traditional Inuit culture, while no similar tradition is found among people living in the Fenno-Scandinavian area.

Hence, within a North Norwegian winter tourism context, dogsledding is no traditional product. Rather, based on the way it is organized, dogsledding constitutes an experience product suitable for pleasure, sightseeing or for soft adventure-seeking people. Studies show that dogsledding is closely associated with the Norwegian winter (Prebensen 2007; Borch 2012) and is also in widespread use in Swedish adventure tourism (Sandell and Fredmann 2010).

This presentation of the Finnmark Race is primarily descriptive and exploratory. It is based on what people involved have told us, and has an emphasis on vital elements, events and changes as they see it. First, the chapter refers to theory

Figure 8.1   Dogs and mushers in the Finnmark Race
Source: Ole Magnus Rapp 2013

concerning the socio-cultural significance of sporting events, focusing particularly on symbiotic relations between the development of culture and industries. Subsequently, the Finnmark Race is described; the start-up, its fashion and forms, the dogs and mushers, the volunteer aspect, and the mediation and mediatization of the race. The development of sled dog tourism is then presented as part of winter tourism, and, finally, the understanding of the process as symbiosis is discussed.

Sporting Events and Festivals – their Significance for Tourism and Community

As events continue to play a more important role in modern society, they also become objects for research. While most studies deal with matters of economic revenue and impact (cf. Burgan and Mules 1992; Crompton et al. 2001), some consider the sociocultural value of events (cf. Dwyer 2001; Mackellar 2007), and others deal with events as catalysts or leverage factors (Chalip 2006b; O'Brian 2007). Most sporting events originating from communities tend to involve a large number of local people and are important for many of these people's social lives (Kvidal et al. 2012). Many have merely a mental involvement, identifying with the event. Others take part as volunteers. According to Botes and van Rensburg (2000), volunteering produces community participation and feelings of unity. It is also maintained that regional events can reduce conflicts and enhance commonalities between communities (Schulenkorf 2009).

Concerning the relationship between sporting events and tourism, authors have characterized it as a 'fit' or 'match' (Florek and Insch 2011), as synergies, (Harrison-Hill and Chalip 2005) or symbioses (MacKellar 2007). The following analysis draws on the latter perspectives. The point of departure is that the dog race and winter tourism not only constitute a 'match' of intertwined trajectories, but also represent an example of a symbiotic relationship with reciprocal gains and synergies. There are traditions within industrial research looking at such examples of symbiosis (Chertow et al. 2008). As will be shown here, the Finnmark Race has provided a foundation for dog-based tourism, while the dog sled tourism sector has in turn provided a base for many race mushers. In most cases, industrial symbioses take place in networks (ibid.) and the Finnmark Race can be viewed as one such network that includes local and national organizations for sled dog enthusiasts.

Sporting events and tourism are tied together in series of ways. First, there is often a direct relation between sport and tourism in terms of an activity that is suited for tourism, sport facilities that function as attractions, and sport infrastructure that can be utilized as tourism infrastructure (Zauhar 2004). Dogsledding exemplifies a sporting activity that has proved suitable as a tourism product; the copying of not only activities but also the race trails and the use of race stories has become a popular way of theming tours.

Secondly, to some extent, sporting events are tourism. A sports competition gathers athletes, fans and other spectators. Additionally, such events are increasingly accompanied by other events or side-events (O'Brian 2007), which enlarges the total effects. According to Ziakas, 'Sociability beyond the venues is enhanced mainly through event related social events. The organization of ancillary events increases the social value of the event portfolio' (Ziakas 2010, 155).

Thirdly, there is normally a symbolic symbiosis in that the sport event creates awareness not only of the sport, but also of the event location as a destination for travellers (Xing and Chalip 2006). Events are often place-branding venues. As part of this, products related to an event tend to be sold under labels and logos. Excursions, tours, souvenirs, gadgets, clothes and other merchandise are part of the event package, thus underpinning the entire event construction and communicating the tourism system of the destination.

Fourthly, there tends to be a strong relationship between sporting events and local and regional development. Tourism is an arena for such development. Sports events are, as Green argues, 'no longer merely about providing good sport. They have become a common tool for local and regional economic development' (Green 2001, 4). Furthermore, she continues, 'organizers are expected to attract as many visitors as possible ... and organizers have had to invent ways to make events appealing to more people' (ibid.). Event organizers have to think beyond their prime values in composing their portfolio. Sporting events are identity markers for the places and regions involved, and important for their development and marketing.

Table 8.1   Interviewees

Interviewee	Role
Sven Engholm	Pioneer in dog sled racing, initiator of the Finnmark Race, and tourism operator
Roger Dahl	Racer, a front figure in the Finnmark Race and dog sled tour operator
Sled dog racer	Racer and tour operator
Arne O. Nilsen	Early sled dog racer and volunteer
Volunteer – local	Media figure
Volunteer	English, repeat visitor
Tor Mikkelsen	Tour operator
Tourism actor	Camp site owner
Tourism actor	DMO person, shareholder and consultant
Kari Jæger	Racer, volunteer, DMO, tour operater, researcher and co-author of this chapter

Methods

This chapter is partly based on interviews with people that have been involved in the development of the Finnmark Race and sled dog tourism in the region. However, the story told is also based on observations and reflections made by the authors, both of whom have lived in the community where the studied event takes place. Furthermore, one of the authors has been part of the development of the sled dog race studied and tourism development. Hence, there are elements of autobiography involved in the presentation to follow. When balancing 'how much of their "self" to inscribe in their "texts"' (Feighery 2006, 273), the intention here is to give a presentation of the development of two intertwined processes seen through the eyes of different stakeholders and parties involved, including one of the authors. Because only one of the interviewers and authors of this chapter has been involved in the race, those informing the reconstructions of the history of race-tourism relations have been challenged from both insider and outsider positions. However, the text presented here is a researcher based narrative.

Among the total ten respondents four are strongly tied to the Finnmark Race, four have been or are in the tourism industry, and three have had positions in both. Two of these have been racers and tour operators combined. Table 8.1 shows the interviewees. In addition, we talked to two volunteers, one investor and authority-based stakeholders, though not within a formal interview setting. These interviews and conversations were all recorded, and have been partly transcribed. Some interviews and articles found on the home pages of the race are also quoted.

The Finnmark Race – its Start and Development

The start-up

The Finnmark Race was organized for the first time in 1981, by Alta Sled Dog Club *(Alta Trekkhundklubb),* which has been in charge of the race ever since. The idea of holding a sled dog race in Finnmark was developed by Sven Engholm, who later became one of the most reputed mushers in Norway. He stands out as a pioneer of sled dog racing and sled dog tourism (see Box 1). He also took part and won the first race in 1981, where three dog teams participated.

Box 1   Sven Engholm – a sled dog and winter tourism pioneer

Sven Engholm is a pioneer, both in dog racing and winter tourism in Finnmark. He was born in Malmø, in the south of Sweden. He moved to Kiruna in Sweden in 1973, where he for some years worked as a social worker and school teacher. He was then already an environmentalist and a dog enthusiast. In his childhood, he read Jack London and used his sister's Afghan greyhounds in front of a sledge. He bought his first polar dog on his way to Kiruna, and in 1978 started together with Christer Johanson a sled dog race called the Nordic Marathon. This was the starting point for long distance sled dog races in Scandinavia.

Later Engholm got a job at Finnmark Leirskole in Øksfjordbotn, Norway, a holiday camp for schoolchildren. They had sled dogs, and dogsledding became part of his job. Here he met his first wife. They moved to Alta, as she was a teacher student there. When she finished her education, they decided to go to Karasjok in 1980 to start a tourism business. They hired a farm, but after some years (in 1985) they got hold of an area where they could develop a dog kennel and establish a permanent tourist camp. Since then Engholm has developed a significant tourism camp on that spot. In 2013, this includes a dog farm, several cabins, a kitchen and a small restaurant, a family house, and rooms for staff – most of it handmade by Engholm himself. He is still an environmentalist and a lifestyle entrepreneur. He characterizes himself and his wife as 'nature hippies' when they started the company back in the early 1980s. To live in and with nature, and in this way perform a good life, is his philosophy.

To be able to live with nature and have a good life, he has not desired for his tourism operations to grow excessively. He does not want to end up as a manager without time to cater for the tourists himself. He is still a musher working with tourists. At the most, there are five people capable of guiding tour groups. The company offers different tours, from four to twelve days. He does not offer day-trips or short trips for conferences or meetings. To accommodate his tourists he uses an old existing network of state owned mountain lodges, but he also has a semi-permanent tent camp of his own. The tours are of high quality and are relatively expensive. The economic condition of the company is good: 80% of the income is from winter tourism activities, but he also offers accommodation in the camp during summer time. The company does all its own marketing and sales work, through tour operators and agents in a few

European countries and through the company's web site, which includes a booking service.

The idea of a Finnmark sled dog race came up when Engholm still lived in Alta. He and a couple of friends wanted to test the dogs, the equipment and themselves as mushers. The story also runs that the dog owners wanted to find out which dogs were the best. As it was, they decided to set up a race between Alta (Sorrisniva) and Karasjok. Three equipages were out together in 1981. Engholm, himself was in one of them and won. One of equipages ended the race shortly after the start – the sledge had not been of a suitable quality. He credits a couple of enthusiasts in Alta for the fact that the race was organized and became a yearly event. Engholm took part in the discussions about how the race should be developed and around the development of dogs, training methods and breeding. He was the only professional in the field at that time, professional in the sense that he worked with dogs on a full-time basis as a sled dog tourism provider. Not surprisingly, his dogs and he as a musher were for many years better trained than his competitors. He won the race eight successive years, eleven times in all. On four occasions, he has also participated in Iditarod, the longest and most famous race on earth, with a best finish of ninth. He was supported by sponsors with passion for the sport.

Engholm was vital for the development of the sport, as he systematically gathered data about his dogs and the races using a computer based system. According to other mushers, Engholm was and is always open to discuss his experiences and knowledge with the others. Together with of a couple of others he worked systematically on creating knowledge about dogs, their health, feeding and breeding capabilities. Engholm emphasizes the importance of the Finnmark Race in the development of the sport. It was where the mushers met and talked; it was the arena for creation of cohesion and a musher culture, concerning the development of knowledge about polar dogs and a particular type of dogsledding. Engholm has never had contact with research institutions, as the veterinarians of other milieus specialized in the domestication of animals. There is no link to formal knowledge institutions in this respect.

Engholm very much stands out as a self-made man. He moved to Karasjok to undertake a quiet nature-based life. This is still his mantra. He is also aware of the fact that he is a model for others starting in the field, and he supports all those asking for advice. Several of his former employees have started sled dog based tourism companies after leaving him. Engholm is part of a small informal network of sled dog mushers, three sled dog companies in Finnmark that meet once a year sharing experiences. He also buys experience products from other local companies, for instance reindeer herders, as visiting a flock of reindeer is one of his offers. However, he has no desire to take on a central position in organizing tourism or dog sledding as an industry.

Several respondents describe internal controversies that have occurred around the strategic direction of the race over the years. One point of dispute was about whether the race should be commercialized. In the beginning, non-commercial interests dominated, but a turning point came in 1989 when celebrity mushers like Susan Butcher and Joe Runyan from Alaska visited Norway. Drawing on their experiences from Alaska and they emphasized the importance of high prize money for the reputation of Iditarod, both in Alaska a across the USA, while highlighting similar potential for the Finnmark Race in Norway and Europe. One of its early participants admits that, with dog racing merely as a hobby, it is hard to survive: 'Sled dog racing is an expensive sport, if you have ambitions to win. You can see from the result list the last years that the top ten to fifteen mushers either enjoyed a strong financial footing or were semi or entirely professional dog mushers'. Therefore, some have developed dogsledding into tourism ventures, while others have found niches for using dogsledding in the social care sector, for instance.

Up to the early 1990s, the race largely relied on a group of volunteer enthusiasts. A number of interviewees said that this was down to passion – passion for dogs, nature, the race, and the event. During the 1990s, the number of participants grew, including more mushers from foreign countries. Volunteers administrating the race used their personal resources and networks to arrange and develop the race. The race was at a crossroad: should it continue to exist as a sled dog race arranged by volunteers, with low budgets, low prizes and remain unknown beyond those in the sled dog milieu? Or should and could it grow economically, increasing value for the mushers, race owners and other stakeholders? A volunteer and tourism industry actor recalls how, in around 2000, she contacted several business people about the challenges facing the race. In 2001 six investors, including Alta Sled Dog Club, raised 100,000 NOK to set up a shareholding company. Since then, this venture has assumed the role of assisting Alta Sled Dog Club, managing the race, including sponsor and marketing relations and media. With the shareholding company in the lead, race strategies took a commercial turn. The company started with one staff member in charge, hiring others on a part-time basis. In 2013, the race administration budget covered four full time jobs and around 600 volunteers were involved in the race.

The fashion and form of the race

The Finnmark Race starts in Alta, on the Saturday of week ten in March each year. In 2013, it was arranged for the 33rd time. At the start, a total of 126 mushers from 13 countries lined up, of which 94 were from Norway, together with 1,350 dogs. The start and finish are positioned in the main street of Alta, with approximately 2000 spectators in attendance. There are two race classes, 500 km and 1000 km. For the 500 km class, it is the real start, while for the 1000 km the first relay is an exhibition, with sponsors and celebrities in the sledges, and with a restart after 15 km.

The length and track of the race have been discussed and changed over the years. The length of the model, Iditarod, is 1800 km. The tour in 1981 was 262 km, the year after it was named the Finnmark Race and enlarged to 400 km, later to 600 km and 1000 km (since 2002). The circuit runs from Alta to Kirkenes and back (see the map, Figure 8.2). After a couple of years of only the 1000 km race, and a fall in participation, organizers introduced two new classes, including a 'short' 500 km. The itinerary has been subject to minor changes, but as time

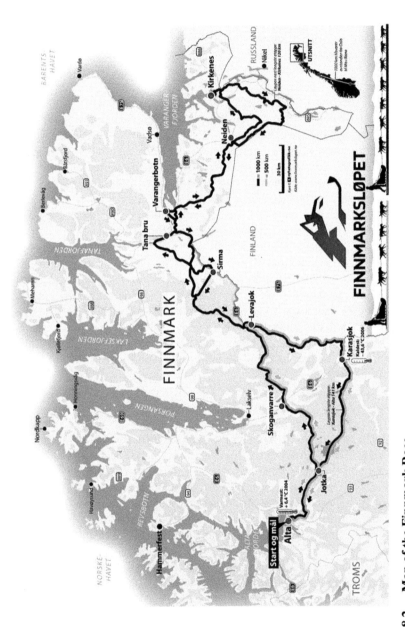

Figure 8.2   Map of the Finnmark Race

Source: The Finnmark Race, web page, 2013

Table 8.2 Aspects of the Finnmark Race, 500 km and 1000 km, 2012 and 2013

	500 Km 2012	500 Km 2013	1000 Km 2012	1000 Km 2013
Mandatory rest hours	20	20	24	24
Number of participating teams	67	70	52	56
Female participants	17 (Winner)	17	7 (Winner)	10
% finishing	83,5	68,5	46	60,7
Prices for winner	NOK 28 000 Polaris ATV (NOK 85 000)	NOK 29 000 Polaris 500 (NOK 89 875)	NOK 64 000 Polaris ATV (NOK 115 000)	NOK 65 000 Polaris 550 (NOK 114 875)
Time spent by the winners	31:16 work 20:43 rest	31:12 work 20/38 rest	88:47 work 56:53 rest	85:47 work 54:16 rest

has passed, the race has become more accessible and tourist friendly as the track follows the main east-west highway in the region (Jæger and Olsen 2005). One municipality has been excluded because of difficult negotiations concerning the itinerary and in order to avoid reindeer grazing areas. Some basic data concerning the race is shown in Table 8.2.

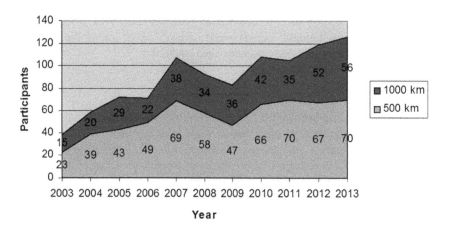

Figure 8.3 Number of participants in separate classes 2003–2013

Figure 8.3 shows the number of mushers over the last ten years. The figure shows an upward trend in the number of participants. The finishing rate varies between 46 and 90%.

The Finnmark Race was started by and for the mushers. For a long period, the organizers were neither interested in looking upon it as a tourism event, nor as a sports event, and most mushers were generally extreme outdoor recreationists or expeditionaries. In recent years, the race has increasingly taken the shape of a sport competition. The times are changing, one of the pioneers said, referring to the support given by handlers to mushers and the fact that most participants are accommodated in caravans at the checkpoints. Moreover, according to several veterans, one no longer a needs to be a trained expeditionary to take part. They perceive many of the 2013 mushers as athletes demanding a streamlined race, not as wilderness people. When the weather is bad (as in 2012), these athletes tend to break the race. In 2013, an experienced musher from Alaska, Hugh Neff, participated. He came in sixth. Neff commented on the fancy campers and logos that most mushers have in the Finnmark Race: 'I don't want that, I just wanna be a dog musher. It's about having fun and still racing. That you do when the dogs are eating good, pooping good, peeing good and are ready to rock'n roll. Let them do what they love' (Finnmarksløpet, web page (Andersen 2013)). Thus, he is critical of the commercialization, sportification[1] athletification and the technification of the race. Other respondents claimed that streamlining reduces the tourism potential; dogs are behind barrages and the mushers are in their caravans – not much to see.

The Finnmark Race has since its start been an important knowledge development arena for the sled dog sport. With less contact and stronger competition, Engholm, the pioneer, raised the question of where these arenas currently (2012) are. In previous years, race mushers met in different settings, socializing and discussing the sport. These meeting places have disappeared because of mobile homes, because of the services mushers are offered along the trail, and because of the mushers' focus on sponsors and commerce. Previously, the race was the most important forum for dialogue and knowledge dissemination among the mushers. But not so much anymore, other agendas have taken over.

Dogs and mushers

According to our respondents, the past is littered with stories about mushers being almost as wild as their dogs, dogs that were fed with raw meat, and about blood flooding in dog fights. Contemporary stories are different. Dog racing used to be a male sport. Now it is a sport where men and women are equal at the starting line. In 2012, there were female winners in both classes, and Inger-Marie Haaland has won the long race twice. Regardless of the reasons for the softening of the

---

1   According to Lippe (1994), This is a process of transforming a physical activity into a modern sport focussing on achievement and competition, undertaken within a rational setting.

Figure 8.4   The Alaskan Husky has become attractive for sled dog activities

Source: Ole Magnus Rapp 2013

image, this transformation may prove an advantage for the tourism marketing role of the race.

The Finnmark Race is referred to as a core arena for the development of the dogsledding sport in Norway. Along with the development of the race, the level of competence on breeding, feeding, taming and training the dogs has grown within this milieu. Part of this is a conscious strategy to get rid of aggressive types of dogs and genes. There also has been a shift from Greenland dogs and Siberian huskies to Alaskan huskies. The latter is not a registered breed, but attractive for sled dog activities because they are bred with healthy physical features and a good temper. Taming and training is based on awards, not on punishment, it is maintained – in most dog teams the dogs are treated like pets. At the start of the 2012 race, a former participant commented on the lack of barking. All this adds to the 'softening' of dogsledding and makes the dogs more suitable for tourism.

Volunteers

In terms of volume, methods, and significance, the level of volunteering in the Finnmark Race constitutes a manifest volunteer tourism. In its first twenty years, the Finnmark Race was organized and run entirely by volunteers. What was rather typical for events and festivals (Monga 2006; Aagedal et al. 2009), was also typical in Finnmark (Jæger and Mykletun 2009). Through volunteering, locals have become tied to the events, and such involvement creates a culture of volunteering (Misener and Mason 2006). Over the years, greater numbers of non-local volunteers have been recruited, including previous residents of Alta returning as volunteers. The volunteers are often the public face of the race during the event and its ambassadors throughout the year. A British volunteer expressed his fascination with Arctic nature while adding that 'I like the people, they are kind and willing to talk with you, and remember you from last year'. This exemplifies the importance of the social bonding aspect of volunteer tourism (McGehee and Santos 2005).

An important group of semi-volunteers perform the role of race veterinarians. Already by 1982, the Finnmark Race had begun collaborating with local vets. According to Norwegian law, vets shall monitor events involving animals (Arctander 2005). The number of vets in 2013 was seventeen plus four vet students and one animal nurse, drawing experience from Italy, France, Canada, USA, England and Norway. They work for free but receive expenses for travel and accommodation. One of the vets says that 'it's about the fascination and the interaction between the dog, the musher and the vet. And the impressive body and physiology the racing dogs have. And of course, it is about the love of nature and of outdoor life' (Finnmarksløpet, web page (Holm 2013)). These people are not only semi-volunteers; they also represent a type of work-related tourism.

Media coverage and public awareness

In its early years, the race did not receive much public or political support, nor did it attract much media attention. Their model, Iditarod, already had a strong media profile. One exception is local media outlets, which have given the race significant coverage since the beginning. One of the interviewees in this study, Arne O. Nilsen, said that he was the 'press officer' for the first race, when he gave an interview to the local newspaper *Finnmark Dagblad* (February 25, 1981). Several volunteers had networks including media representatives, from whom they learned how to provide media support, and in particular how to organize a press centre and sponsorship for television companies. The press centre had, for a long period, a 'permanent' core of volunteer staff. One volunteer from the 1990s said that she carpet bombed the media for years. Some had asked her why she was sending out all the information – nobody came anyway. Although her activities produced minor results, she noticed that the mushers' local newspapers often used the information. This gave the race, and Finnmark more generally, some important media coverage. The first time national media showed interest was in 1993 when a hurricane occurred during the race and many mushers had to lie down to wait for better weather. In their Daily Review, a national television channel reported on the 'missing' participants. In 2001, media coverage of the race got a lift-up. The race was asked by the county governor to make a one-day program for the Norwegian Princess Märtha Louise. Her visit received much attention, and for the first time media were accredited. Since then, the Crown Prince has also attended the event. More recently, the participation of Lars Monsen, a national celebrity known for wilderness television documentaries, brought about an increase in media coverage.

**Figure 8.5   Celebrities at the start: Crown Prince Håkon, and musher
Lars Monsen, a wilderness documentary producer**
Source: Ole Magnus Rapp 2013

The use of celebrities is a well-known marketing technique (van der Veen 2008) and is said to be particularly effective when endorsers have appearances congruent with the brand (Misra and Beatty 1990). Excepting the role of celebrities, the commercialization and professionalization turn seems to explain the growing media interest in the Finnmark Race.

The most important shift came in 2009, when the national television broadcaster NRK started to make daily evening programs focusing on the race and the surrounding folklore. In the first few years, this production was sponsored. In 2001, forty-eight journalists were accredited; and about 130 in 2013, of which forty represented NRK. Since 2012, every sled has had a GPS tracker attached, allowing interested parties to follow the mushers on the internet. This web interest is significant. During the week of the 2003 race, there were 1.4 million internet hits without trackers on the teams. In 2011, the leading mushers had GPS trackers, resulting in 57 million hits. In 2012, with trackers on all mushers, there were 111 million hits, and in 2013, 143 million hits. Throughout, the race, sled dog tourism, and Finnmark as a destination have received significant attention.

Media coverage of the Finnmark Race is a telling indication of its position both within the dog sport community and in society as a whole. Television and internet exposure stories are filled with exotic pictures from polar environments. Its media success can also be explained by the values the race conveys and the culture it represents. Although not a strong part of it, dogsledding fits into the

Norwegian outdoor recreation culture. Importantly, it is a reminder of a heroic history and of national pride, a celebration of Nansen and Amundsen and their expeditions, and the heroes from the 'race' in Alaska in 1925. It also fits into a television genre that is extremely popular in Norway – wilderness expedition documentaries. This exposure encourages local people to get involved. Local involvement tends to increase when people, through a televised event, can watch their home place on television (Kotnik 2007). This also seems to have advanced the folklore aspect of the event. There are now several festivals, shows and other events that gather spectators at checkpoints, as in Tana and Varangerbotn. This adds to the involvement factor. As Grohs and colleagues show (2004), there is a strong correlation between high levels of involvement with an event and high levels of awareness of sponsored products. The respondents comment on this matter as strengthening the race as a platform for tourism.

The Winter Tourism Start-up in Finnmark

For a couple of decades, sled dog tourism has been part of winter tourism development in Finnmark. Winter tourism in the region has never been big, but the Arctic winter, the northern lights, reindeer husbandry and Sami culture have always attracted visitors. For many decades, skiers have crossed the Finnmark Plateau (*Finnmarksvidda*). All over Finnmark there is an infrastructure for travellers dating back more than a century, including roads, marked tracks, and government run guesthouses (*Statens fjellstuer*). This is also an infrastructure for hikers, skiers, and more recently for cyclists. Following the tracks, it is normally not more than 30 km to the nearest lodge. Thus, an infrastructure that existed when sled dog tourism began is just as important today. It started around 1980 with one company with a few customers, while in 2013 the number of companies is close to twenty, including part-time ventures. As the products vary from one hour to one week, the number of customers is not a good measure of success. The six biggest companies had a combined total of 9,800 customers in 2013, an increase of 9% from 2012, but a company offering short trips to Hurtigruten accounts for over 5,500 of these tourists.

As mentioned, Sven Engholm is a sled dog tourism pioneer. Another is Roger Dahl, a musher of fine repute. Dahl had moved to Alta because of the dog milieu and, in 1983, he took part in the race for the first time. After a few years combining tour operations with an ordinary job, he became a full-time tour operator. Engholm primarily offered the expedition type of tours, whereas Dahl also facilitated day visitors. Due to the perception that the market was small, other mushers did not go into this business out of concern that they would spoil the ground for the two that had started. Later, several others joined the business. Today, there are about eight full-time sled dog tourism operators in Finnmark, and about five operating on a part-time basis. All eight full-time tour operators have taken part in the Finnmark Race. Table 8.3 shows the situation in 2013: of

Table 8.3   Number of mushers in the Finnmark Race 2013, which have their own tourism company

Race Class	Finnmark	North-Norway	Rest of Norway	Foreign	All together
500 KM	2	2		8	12
1000 KM	4	1	2	4	11

all the participants, twenty-three run a tourism company. In addition, some of the race mushers are employed by such ventures.

It gives a certain authority as a tour operator and guide, it is said, to have taken part in the Finnmark Race. The stories are better and the promotion is more convincing. Tourists who have visited a race kennel tend to follow the musher on the internet during the race in subsequent years, we are told. However, and despite a professionalization process that can be observed, most mushers retain dog sledding as a hobby and spend part of their spare time racing.

Concerning tourism related to the race, Tor Mikkelsen stands out as a pioneer. He started as a tourist guide in Alta in the 1980s and built up a tour operation company. With a growing sled dog milieu in Alta, people had on several occasions asked whether this could somehow act as a platform for developing tourism. Together with a colleague, he created a project on sled dog tourism, for which he received public funding. For some years, Mikkelsen offered package tours following the race, but this was never a big success. Many mushers lined up against this development, regarding Mikkelsen and the tourists as 'parasites'. This attitude has changed; today, such initiatives are strongly welcomed. Mikkelsen was a kind of mentor to aspiring tourism operators and had a constant focus on quality.

Sled dog tourism has found both a competitor and a partner in snowmobile tourism. The start-up period was the same and Tor Mikkelsen had a role in developing both tourism ventures. He saw them as complementary in the development of winter tourism product from an early stage. To these two components was added a third: exposures and activities related to Sami culture (visit to a tent (*lavvu*), reindeer sledding, Sami food, music (*yoik*), and narratives). A fourth component was later added when one of the operators of snowmobile tourism went into the ice-hotel business (*Sorrisniva*). The fifth component is northern lights tourism – often combined with dog sled tours – which has become particularly popular during the last few years.

Concerning winter tourism development in Finnmark, the interviewees were asked to identify the major driving forces. They admitted that there have been ups and downs, with much discordance and envy, and that several major companies have seen the others only as competitors, exhibiting reluctance to cooperate. There is an accordance concerning the important role that some entrepreneurs – Sven Engholm, Roger Dahl and Tor Mikkelsen, for instance – have had for in winter tourism development. Important also has been the snowmobile and ice-hotel company (Sorrisniva), which has recently (2013) become one of the sponsors of the Finnmark Race. Similarly, in the 1990s, staff from the major hotel in Alta had important roles. A regional winter tourism project that was started in 2002 is also said to have had positive effects. This was a strategic project in which goals were set, and experience products scheduled, coordinated, and marketed for individual tourists. Furthermore, the Hurtigruten, the coastal steamer, was mentioned. Since 2006–2007, this company has worked strategically with winter tourism. Dog sled excursions in Tromsø and Kirkenes are among their most popular products. Those who have managed to establish winter cruise ship tourism in Alta also hold some importance for the dog sled companies. One owner claims that the expansion of his company would not have been possible without these people. Concerning the Finnmark Race, the respondents perceive it as a catalyst with particular importance for branding and image building.

As mentioned, the Finnmark Race can also be seen as event tourism. Even if they are few, several of the respondents provide a useful perspective on such tourists coming on individual basis as well as in small groups of package tourists. This will probably expand from 2013 onwards, as there are negotiations going on concerning cruise visits to several checkpoints. In recent years, conferences and meetings have also been held in the race week. The remaining tourists are somehow involved with the race in this capacity or as business tourists. Firstly, there are mushers, their teams, sponsors and guests. The mushers, handlers, family and friends constitute significant groups around each team. With 126 teams, this amounts to a lot of people. For sponsors, the race arenas are sites for meetings and promotional events. Secondly, producers and dealers of equipment and other merchandise attend. Thirdly, the local business community invites customers and holds meetings with partners during the week. Fourthly, it is claimed that the media accounts for more than130 of the total number of people registered. Altogether, these groups constitute a substantial amount of people, representing significant business for the local tourism industry. The hotels and campsites have a good period, while the event is also important for experience producers. The dogsledding tourism providers make money, as do retail businesses, for instance those renting snowmobiles and ATVs.

In terms of cooperation between the Finnmark Race and the tourism industry, it has thus far been restricted to two hotel chains, transport companies and a number of campsites. In 2013 Sorrisniva, the ice-hotel and snowmobile company, became one of the Finnmark Race's main cooperating partners. 'This is a very good deal for both parties', said the general manager of the Finnmark Race. 'Sorrisniva is a leader in adventure tourism, bringing something new into our team. The Finnmark Race is working with a closer link to the tourism industry in terms of further development of the event and therefore this is an important step in the right direction', the general manager adds. Sorrisniva wants to help strengthen Finnmark's position, both nationally and internationally. According to their marketing director, the race is a very good venue for marketing their products: 'We have had the pleasure of accommodating the restart for many years already, and it's high time to formalize cooperation to get more out of it for both parties' (Finnmarksløpet, web page, 2012).

Concluding Remarks

To sum up, the main factors in the success of the Finnmark Race seem to have been hard work, stubbornness, patience, passion, a local anchoring, and a national interest in the event. Currently (2013), the Finnmark Race is the most well-known and best reputed sled dog race in northern Europe, and a good marketer of sled dog tourism. None of its successors or competitors in neighbour countries have reached the same position. There are probably several reasons. One is its polar location and another is the benchmark, the Iditarod in Alaska. By somehow living up to this event, the Finnmark Race has achieved a profile as 'tough'. Another reason is its image, which, as argued here, is related to history of Norwegian polar exploration and the national outdoor culture. Thirdly, as emphasized by several respondents, the sled dog people have strong positions in the national dog race organization, which also is a member of the Norwegian Sports Association (*Norges Idrettsforbund*). Thus, the race is a well organized event within a serious sport context. A fourth factor is its commercial side, with sponsors and the significant prizes serving as great incentives for participation. The media exposure the race has attained has increased its popularity not only in the public, but also among mushers. Lastly and most important, sled dog tourism has demonstrated that it is a viable industry, capable of giving a livelihood to many mushers.

Dog racing and sled dog tourism are obviously good fits or matches (see Florek and Insch 2011). It has been an intertwined development, primarily due to the overlap of people. Indeed, the pioneers of the Finnmark Race and the pioneers of dog sled tourism are very much the same. They clearly represent the interdependence of these two sorts of accomplishment. Preparing for the race is a way of developing the potential for tourism; development of competence, accumulating dogs and gear, creating a financial platform for a sled dog related lifestyle. In the footsteps of the pioneers, there are a whole series of people combining racing and tourism. There is a symbiosis between the activities.

There is also a symbiosis on a societal level. This study has not assessed the impacts of the race, yet they are obviously significant. Likewise, the race has been, and continues to be, a significant leverage factor (Chalip 2006 a, b). It has led to dog sled tourism, media coverage of the region, and a growing awareness of and interest in the region. However, there are also processes going the other way, processes energizing the race. The business community uses the race as an arena for meetings, conferences, and as an opportunity for sponsorship and marketing. This has strengthened the race significantly. The event-business sector symbiosis matches theories about industrial symbiosis (Chertow 2007). Such symbioses tend *not* to be based on collaboration, common strategies or 'from highly structured planning processes' (ibid., 26). Such structures are more like eco-systems, where the system effects are emerging from geographical proximity and 'opportunistic business decision[s]' (ibid.). This is also an accurate representation of the Finnmark pattern. However, on both their parts, the race and winter tourist destination development are based on a number of strategic considerations and negotiations.

The most important processes in the Finnmark Race are commercialization, volunteering, technologization, mediatization, and branding. There are certainly also other important aspects not under focus, for instance festival and identity related issues. An important outcome of the race and of sled dog tourism is an ameliorated position of Finnmark as a winter tourism destination. However, there is also a paradox therein. The growth of the race is primarily caused by the recruitment of mushers from other regions, many of which are involved in sled dog tourism. Thus, a challenge for the region is that there are competing destinations and companies all over the country. In downhill ski tourism, it is shown that additional products were the most vital selling points (Tuppen 2000). This is Finnmark's chance. With the Finnmark Race in the lead, and products related to 'the Arctic', 'polar' location on the Northern Rim, the northern lights, and the Sami culture, the region has unique advantages that place dog sled tourism in the region in a strong position.

References

Aagedal, O., Egeland, H. and Villa, M. (2009), *Lokalt kulturliv i endring* (Oslo: Norsk kulturråd).

Andersen, T.A. (2013), 'I just wanna race, man!', interview with Hugh Neff, *Finnmarksløpet* <http://www.finnmarkslopet.no/article. jsp?id=6268&lang=no.> (home page) accessed 31 March 2013.

Arctander, S.M. (2005), *The Finnmark Race. The World's Northernmost Sled Dog Race* (Røyse: Norsk Bokforlag).

Borch, T. (2012), 'Vinterturisme i den nordlige periferi – fra potensial til suksess i Troms', in Forbord et al. (eds).

Botes, L. and van Rensburg, D. (2000), 'Community Participation in Development: Nine Plagues and Twelve Commandments', *Community Development Journal* 35:1, 41–58.

Burgan, B. and Mules, T. (1992), 'Economic Impact of Sporting Events', *Annals of Tourism Research* 19:4, 700–710.

Chalip, L. (2006a), *The Buzz of Big Events: Is it Worth Bottling.* Research Note, Centre for Leisure Manangement Research, Bowater School of Management and Marketing (Geelong: Deekin University).

—— (2006b), 'Towards a Social Leverage of Sports Events', *Journal of Sport & Tourism* 11:2, 109–127.

Chertow, M.R. (2007), '"Uncovering" Industrial Symbiosis', *Journal of Industrial Ecology* 11:1, 11–30.

Chertow, M.R., Ashton, W.S. and Espinosa, J.C. (2008), 'Industrial Symbiosis in Puerto Rico: Environmentally Related Agglomeration Economies', *Regional Studies* 42:10, 1299–1312.

Crompton, J.L., Lee, S. and Shuster, T.J. (2001), 'A Guide for Undertaking Economic Impact Studies: The Springfest Example', *Journal of Travel Research* 40:1, 79–87.

Dwyer, L., Mellor, M., Mistilis, N. and Mules, T. (2001), 'Forecasting the Economic Impacts of Events and Conventions', *Event Management* 6:3, 191–204.

Feighery, W. (2006), 'Reflexivity and Tourism Research: Telling and (Other) Story', *Current Issues in Tourism* 9:3, 269–282.

Finnmarksløpet (2013) 'Map of The Finnmark Race' *Finnmarksløpet* <http://www.finnmarkslopet.no/page.jsp?ref=fl-map&lang=no> (home page), accessed 29 March 2013.

Florek, M. and Insch, A. (2011), When Fit Matters: Leveraging Destination and Event Image Congruence', *Journal of Hospitality Marketing & Management* 2:3/4, 265–286.

Forbord, M. et al. (eds) (2011), *Turisme i distriktene* (Trondheim: Tapir Akademisk Forlag).

Gibson, H. (1998), 'Sport Tourism: A Critical Analysis of Research', *Sports Management Review* 1:1, 45–76.

Green, B.C. (2001), 'Leveraging Subculture and Identity to Promote Sport Events', *Sport Management Review* 4:1, 1–20.

Grohs, R., Wagner, U. and Vsetecka, S. (2004), 'Assessing the Effectiveness of Sport Sponsorships. An Empirical Examination', *Schmalenbach Business Review* 56:2, 119–138.

Harrison-Hill, T. and Chalip, L. (2005), 'Marketing Sport Tourism: Creating Synergy between Sport and Destination' *Sport in Society* 8:2, 300–320.

Holm, A.O. (2013), 'No Man`s Land', chronicle, *Finnmarksløpet* http://www.finnmarkslopet.no/article.jsp?id=6366&lang=no (home page), accessed 1 April 2013.

Jæger, K, and Olsen, K. (2005), 'Working Holiday? A Race as a Place to Slow Down?', in Kylänen (ed.).

Kotnik, V. (2007), 'Skiing Nation: Towards an Anthropology of Slovenia's National Sport', *Studies in Ethnicity and Nationalism* 7:2, 56–78.

Kvidal, T. Jæger, K. and Viken, A. (2012), *Festivaler i Nord-Norge: Et samhandlingsfelt for reiseliv, øvrig næring, kultur og sosialt liv.* Report No. 12 (Alta: Norut Finnmark).

Kylänen, M. (ed.) (2005), *Articles on Experiences 2* (Rovaniemi: University of Lapland Press).

Lippe, G. von (1994), 'Handball, Gender and Sportification of Body-Cultures: 1900-40' *Journal of International Review for the Sociology of Sport* 29, 2, 211–234.

MacKellar, J. (2007), 'Conventions, Festivals, and Tourism: Exploring the Network that Binds', *Journal of Convention & Event Tourism* 8:2, 45–56.

McGehee, N.G. and Santos, C. A. (2005), Social Change, Discourse and Volunteer Tourism. *Annals of Tourism Research* 32:3, 760–779.

Misener, L. and Mason, D. S. (2006), 'Creating Community Networks: Can Sporting Events offer Meaningful Sources of Social Capital?', *Managing Leisure* 11:1, 39–56.

Misra, S. and Beatty, S.E. (1990), 'Celebrity Spokes-Person and Brand Congruence: An Assessment of Recall and Affect', *Journal of Business Research* 21:2, 159–173.

Monga, M. (2006), 'Measuring Motivation to Volunteer for Special Events', *Event Management* 10:1, 47–61.

O'Brian, D. (2007), 'Points of Leverage: Maximizing Host Community Benefit from a Regional Surfing Festival', *European Sport Management Quarterly* 7:2, 141–165.

Prebensen, N. (2007), 'Exploring Tourists' Images of a Distant Destination', *Tourism Management* 28:3, 247–756.

Sandell, K. and Fredmann, P. (2010), 'The Right of Public Access – Opportunity or Obstacle for Nature Tourism in Sweden?', *Scandinavian Journal of Hospitality and Tourism* 10:3, 291–309.

Schulenkorf, N. (2009), 'An ex ante Framework for the Strategic Study of Social Utility of Sport Events', *Tourism and Hospitality Research* 9:2, 120–131.

Tuppen, J. (2000), 'Restructuring Winter Sports Resorts', *International Journal of Tourism Research* 2:5, 327–344.

Veen van der, R. (2008), 'Analysis of the Implementation of Celebrity Endorsement as a Destination Marketing Instrument', *Journal of Travel & Tourism Marketing* 24:2/3, 213–222.

Zauhar, J. (2004) 'Historical Perspectives of Sports Tourism', *Journal of Sports Tourism* 9:1, 5–101.

Ziakas, V. (2010), 'Understanding an Event Portfolio: The Uncovering of Interrelationships, Synergies, and Leveraging Opportunities', *Journal of Policy Research in Tourism, Leisure & Events* 2:2, 144–164.

Xing, X. and Chalip, L. (2006), 'Effects of Hosting a Sport Event on Destination Brand: A Test of Co-branding and Match-up Models', *Sport Management Review* 9:1, 49–78.

PART III
Reorienting Destinations

Chapter 9

Integrated Tourism Development? When Places of the Ordinary Are Transformed to Destinations

Anniken Førde

Introduction

> The spectacular island of Andøya awaits you with her jagged mountains, midnight sun, giant sperm whales and some of Norway's finest walking … (www.andoytourist.no)

> 'Welcome to the Arctic in North Norwegian Lapland. (…) At Snowman area you are in the Arctic with northern lights, midnight sun and exotic nature, yet still in a modern society with a high standard of living. (www.visitsnowman.no)

These are excerpts of how Andøy and Målselv are presented as destinations. Andøya is presented as 'the magic island'[1] and Målselv as 'a fresh alpine destination in the Arctic'[2] and part of the 'Snowman area'. But these spectacular and exotic places are also places of the ordinary – where people live their everyday lives. The theme of this chapter is how these worlds are related. Through case studies from Andøy and Målselv I aim to investigate how destination development enters into complex processes of local development. A relational approach to tourism development is outlined to grasp how tourism and community development are interrelated through manifold relations and practices. By performing tourism people encounter the culture of others. At the same time tourism becomes part of culture. As a set of social and cultural practices, tourism is both an expression and an experience of culture (Burns and Novelli 2006; Robinson and Smith 2006). Tourism also implies contested culture. The promotion of culture legitimizes and normalizes parts of local realities. Destinations can be seen as a negotiated reality (Ringer 1998), with different actors bringing in different histories, values and perceptions. Tourism development thus involves complex relations of power and identity. Saarinen (1998; 2004) shows that there are several – and sometimes competing and conflicting – discourses on regions and development. Different actors might have

---

1 andoykommune.no
2 malselvfjellandsby.no

conflicting perceptions of landscapes of the north. As the destination of the tourist becomes inseparable from the inhabited landscape of local culture (Ingold 1994), it is urged that a more appropriate tourism development is needed, taking into account the multiple discourses of local development.

Like many municipalities in Northern Norway, Andøy and Målselv are experiencing great changes in their socio-cultural as well as material landscapes. Andøya is situated in Vesterålen, on the edge of the Northern Norwegian coast. The nature is fascinating, with steep mountains, lakes, flat mash-land, white beaches, the endless ocean and continuously changing lights. Andenes is the service centre, and the island has many small fishing and farming villages. Andøya has a rocket range, numerous military installations and an airport, both military and civil. In summer 2012 the road along the west coast of the island was declared a National Tourist Road[3] with a ferry connection to Senja. This landscape also encompasses about 5,100 inhabitants living their lives there, and in summer an increasing number of camping cars and tourist buses. Målselv is situated inland in Troms, between the towns Narvik and Tromsø. Here, too, it is easy to get enthused by the landscape of farmland, pastures, huge wooded valleys and high mountains, with the silvery river and waterfall of Målselva running through it. Bardufoss is the regional centre, and there are many villages. Målselv has a national park and large military areas. E6, the main road through Northern Norway, runs through the municipality, and there is also a military-civil airport. Målselv has about 6,600 inhabitants, many motoring tourists, sport fishermen and hikers in summer, and an increasing number of ski tourists during winter. Traditionally, fishery was the most important industry in Andøy and farming in Målselv. During the Cold War, both municipalities became important sites for military activities, due to their strategic positions. With the new political situation, since the fall of the Soviet Union, these areas partly lost their vital position. In both communities they talk about 'the troublesome peace'; the withdrawal of the military forces resulted in dramatic population decline. As a response to this, tourism is given priority in local development strategies. Andøya is surrounded by an ocean rich with important fish stocks, whale and sea bird populations, and Whale Safari is the main tourist attraction. Målselv has a tradition of attracting people for salmon fishing in the river and hiking in the national park, and in latter years has become an alpine destination through the ski resort *Målselv Fjellandsby*[4] (the establishment of the ski resort is described by Viken in Chapter 7 of this volume). The question to be addressed here is how the creation of tourist landscapes in Andøy and Målselv mesh with the inhabited landscape of local experience, and how tourism development is integrated into broader processes of community development.

In the next sections, the theoretical and methodological implications of a relational approach to tourism development are elaborated. Then, tourism development in Andøy and Målselv is discussed, focusing on the relations

---

3   http://www.nasjonaleturistveger.no/no/andoya.

4   Translates to *Målselv Mountain Village*.

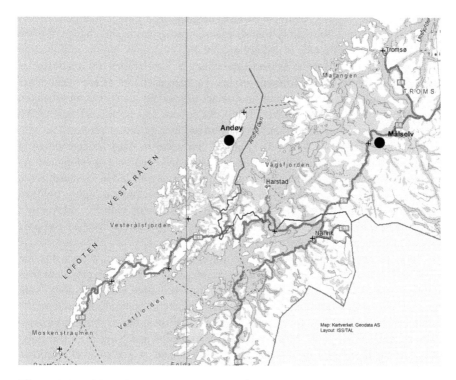

Figure 9.1   Map showing Andøy and Målselv

Source: Statens Kartverk

between tourism and community development. Through analyses of encounters, discourses of development and negotiations of local identities, I aim to explore how tourism development enters into complex processes of negotiating cultural and material landscapes. It is argued that in order to understand how tourism can contribute to developing viable communities, we need to investigate these complex relations between the creation of destinations and broader processes of community development; between tourism landscapes and the inhabited landscapes of local experience.

A Relational Approach to Tourism Development

Tourism as networking practices

I take as a point of departure the fact that tourism plays an active role in creating local geographies, and that these geographies are multiple, dynamic and continuously negotiated. Ringer (1998, 7) states: 'As part of our consciousness and as part of our lived realities, tourism is essentially about the creation and reconstruction of geographic landscapes as distinctive tourist destinations through manipulations of history and culture'. Tourism is both an agent of preservation and change; it often gives an economic rationale to preserving specific landscapes, but the introduction of tourists inevitably results in landscape changes. Landscapes tell many stories and might be contested by different groups. Tourism destinations are then neither an expression of local culture nor an artefact of tourism alone, but an amalgam of multiplicity of influences (Wall 1996). As culture and people are part of the tourism product, tourism cannot simply be read as a business proposition with a series of impacts. Macleod (2011) argues that tourism must be considered in relation to broader social, cultural, environmental and economic development. As places and regions are being planned and transformed in order to attract more tourists and non-local investors in tourism, tourism becomes a highly political and socio-cultural activity (Saarinen 2004). In this study of Målselv and Andøy I look into how tourism as a strategy of community development involves complex processes of negotiating local geographies.

Tourism development implies a commoditization of place; places that are inhabited by different actors with different interests. Tourism development thus implies struggles over representations, meanings, practices and materiality, both between and within destinations. But tourism research has barely focused on the relation between tourism development and broader local development processes (Tinsley and Lynch 2008; MacLellan 2011). As Burns and Novelli (2006) point out, tourism literature rarely acknowledges the world as a system of relations. Integrated tourism has been introduced as a concept to address the complex interrelation between tourism and community development. Butler (1999) defined integrated tourism planning and development as a process of introducing tourism into an area in a manner in which it, in an appropriate way, mixes with existing elements. Further elaborated by Saxena et al. (2007), integrated tourism is seen as tourism explicitly linked to the economic, social, cultural, natural and human structures of the localities in which it takes place. It is assumed that integrated tourism development can contribute to a more resilient tourism industry as well as the creation of viable communities (Saxena et al. 2007). They argue that we must focus on the connectivity between different actors, activities and resources and the diversity of values related to tourism development. In line with Bærenholdt et al. (2004), tourist places are here understood as produced in complex relations involving multiple types of networking practices, producing places as material natures, social relations and discursive conceptions. Through empirical studies of such networking practices it is possible to explore how tourism and destination development is integrated in broader discourses of community development, and how local sensibilities are incorporated in the understanding of tourism development.

A tourist destination is not something that is, but something that becomes (Frisk 1999). As Saarinen (2004) argues, destinations are not stable, closed systems, but dynamic, historical units with specific identities, constantly produced and reproduced through complex social and discursive practices. They are socio-cultural constructed spaces that actively shape their own future. Representations used in destination marketing are not value-free expressions of place identity, but the culmination of historical, economic, social and political processes (Saraniemi and Kylänen 2011). The produced images of destinations are partly shaped through policies as an element of community building (Bærenholdt et al. 2004). Being designed to project an appealing image, tourist destinations to some extent become representations of what tourists seek. As will be discussed in the further analyses, the process of describing destinations influences both tourists' and locals' perceptions of specific landscapes. Tourism thus constantly (re)produces representations of place through processes where spatial boundaries are marked out and meanings of place and who belong in it are confirmed. Local social life is dramatized and cultivated in what Picard (2011) calls 'human gardens'. In the process of tourism development some aspects of culture are privileged and others excluded. Representations freeze living forms and depict reality in incomplete ways, and culture becomes abstracted from social practice. Images of place tend to be the result of stereotyping, where specific aspects of local landscapes are exaggerated, and others understated. Hughes (1998) argues that the ways places are represented reflect an unstable outcome of struggle between interest groups – where the interventions of tourist entrepreneurs, travel writers, governments and tourists have come to play an increasing role. In the case studies from Målselv and Andøy the struggles involved in tourism development are emphasized.

Saraniemi and Kylänen (2011) present a cultural critique of common ways of understanding destinations – through economic, marketing or customer-oriented approaches. They argue that we need to shift the focus from destinations as products to destinations as processes, as they are under constant negotiation and renegotiation that connect to wider discursive frameworks. But destinations not only relate to discursive practices. Applying actor-network theory to tourism, van der Duim, Ren and Jóhannesson (2012) argue for a greater sensibility towards materiality. Talking of *tourismscapes*, complex processes of ordering 'in which a variety of practices, discourses, artefacts and technologies come into play' (van der Duim et al. 2012, 26), they draw our attention to how destinations result from people and things interacting and networking. Taking these critiques seriously, my approach is to address the complex relations involved in the continuous construction of tourism landscapes. The concept landscape is somewhat ambiguous, but as Ringer (1998) argues, it helps to draw attention to the interrelations of material structures and emotional attachment to places, both by inhabitants and visitors. And to the way it is imagined, produced, contested and enforced.

Methodology: Entering into the negotiation of place and development

The analyses are based on case studies in Målselv and Andøy. During 2011–2012 teams of researchers[5] conducted individual interviews and focus group sessions with people involved in tourism development and community life in different ways. We talked with people from the tourism industry, from other industries in the communities, from the voluntary sector, from the local government administration, leaders of the local council, groups of retired people and of young people. Through such a wide selection of informants, we aspired to approach the topic of tourist development from different actors' perspectives and to acknowledge the complexity of relations and networks involved (Berg et al. 2013). Such a multi-sited fieldwork (Kramvig 2007) implies expanding the context of tourism development, and challenges implicit conceptions of relevant relations and coherences. Working with several researchers with different backgrounds together in the field also contributed to expanding the analyses, and make explicit our interpretations and constructions (Kramvig 2007).

The emphasis in the interviews and focus group sessions was on community development in general and tourism development in particular, with a special focus on the multitude of networking relations. As Viken (2013) argues, using focus groups seems fruitful in studies of restructuring processes. Through gathering people in focus groups, common constructions and many dividing perceptions become explicit. In these *ad hoc* group conversations understandings of place and development were negotiated, revealing many shared stories and conflicting opinions. In addition, I have participated in several meetings for tourist enterprises in Andøy, where challenges, possibilities and cooperation have been debated. The two case studies are separate but allied.

As Law points out, our methods are performative: 'They not only describe but also help to produce the reality that they understand' (Law 2004, 5). Through our methods specific realities are made real, at the cost of other versions. Our way of doing research is thus given a creative role in constructing specific understandings and realities (Jóhannesson et al. 2012). Making use of focus groups as a method of examining on-going processes, we also contribute to these processes (Viken 2013). Creating and participating in arenas for discussion, the research team entered into the negotiation of local realities of community and tourism development. And as the negotiations continued after we left Målselv and Andøy, these landscapes constantly change. The account given here relates to the specific time and space of our fieldwork.

---

5   The research team in Målselv consisted of Per Kåre Jacobsen, Arvid Viken, Torill Nyseth and Anniken Førde (all UiT The Arctic University of Norway), and in Andøy of Arvid Viken and Anniken Førde.

Tourism as a Rescue for Restructuring Municipalities

Tourism is seen as a major alternative to the rural decline in many European countries (McCabe and Marson 2006; Saxena et al. 2007; Førde 2011). This has also been the case in Andøy and Målselv. The decline of traditional industries, like fishery and farming, and the dramatic downscaling of the Norwegian Armed Forces in recent decades, has caused loss of workplaces and depopulation. About 1,500 soldiers and 3–400 employees left Andøya, and the population dropped from 9,000 to about 5,200 in a few decades. In Målselv they lost about 500 jobs and many soldiers left, and the population dropped from 8,500 to about 6,600. As a result, Målselv and Andøy were defined as stagnating municipalities with a need to reform their industrial and economic structure, and they received extraordinary allocations from the state. Searching for alternative futures, tourism came up as an important priority.

In their plan of action for business development the municipality of Andøy highlights fishery and tourism as areas of priority. The tourism industry encompasses about 30 tourism enterprises. The most important, according to size, is Whale Safari, with about 12,000–15,000 visitors a year. The Whale Safari was established in 1989, and is seen by many as the backbone of tourism in Andøy. Other tourism enterprises are a couple of hotels, a few restaurants and cafés, camping sites, rooms and cabins for rent, art galleries and shops, sea and bird safaris and activities like kayaking, guided walks and farm visits. Most of the enterprises are quite small, often family based and seasonal. The main tourist season is summer, but winter tourism based round the northern lights is growing.

In Målselv as well tourism has become an area of priority for the municipality. Målselv has long traditions of tourism based on salmon angling in the river Målselva, and the waterfall Målselvfossen can document about 90,000 visitors a year. Målselv is also well known among hunters, hikers, skiers and biathletes. In 2008 a ski resort, *Målselv fjellandsby*, was opened, including ski slopes, cottages and flats for sale and rent and an amusement park, *Blånisseland*[6]. In addition, a destination company, Visit Snowman, was established. Since it started, the development of the ski resort has been a major priority for the municipality. But the tourism industry also consists of a hotel, some guesthouses, a camping ground and several small-scale enterprises offering accommodation, dog sledging, reindeer sledging, horse-riding and guided tours. With different enterprises having different seasons, Målselv has both summer and winter tourism.

In both Andøy and Målselv, the critical situation of demilitarization forms an important framework for understanding the increased emphasis on tourism development. The former mayor of Målselv explains: 'We lost 1,500–1,600 people. So we had to do something to create jobs, to get people back'. According to him, the realization of the ski-resort and connected activities is a result of the

---

6   *Blue Elves Land*, where people can meet the mythical blue elves known from Norwegian children's movies.

thorough work of readapting a military community. This process has involved a range of local and non-local actors, and unconventional alliances of investors, local governments, enterprises, landowners and herders (Nyseth and Viken, forthcoming). In Andøya they are in the middle of a process of establishing more collaborative networks among tourism enterprises. In recent years the infrastructure for travellers has been strengthened: a ferry connection to Senja during summer, direct flights from and to Oslo, and selection as a National Tourist Road. In order to build on the potential this gives for tourism development, many see the need for new networks and alliances. As the secure situation of being a military community has disappeared, they are forced to look beyond established structures and models of action and interaction. In both municipalities, the attention on tourism development seems to be important not just as an employment alternative, but also symbolically. The effort to establish a ski resort in Målselv has gained a lot of media attention. People from the local authorities claim this has changed the focus of the community from depression to concentrating on potential and development.

Even though there seems to be a general acceptance of tourism representing a potential rescue for these communities, the role of tourism as a priority area in municipal policy is contested. Whose interests are being included in such strategies? Are the interests of the large tourism enterprises included at the expense of all the small ones? What about the interests of other enterprises, the voluntary sector and common citizens? These questions were more or less explicitly debated in many of the focus groups, as well as in the meetings on tourism development. How tourism should be pursued is also contested: what tools do the municipality have, and how should they be used? How should the community be profiled, and who are the best people to do so? Tourism development is thus not an easy rescue; the construction of tourism landscapes set in motion a series of struggles over discourses and materiality.

Encounters

Tourism enters into the landscape of local experiences in multifarious ways (Picard 2011; van der Duim et al. 2012). But most of the people we have talked with in Andøy and Målselv said that they don't feel they live in a touristic place, and that tourism does not influence their everyday life-practices much. Many describe a lack of encounters; they don't interact with the tourists. In Andøy an elderly man describes it as some kind of mutual bird watching: 'It's amusing when they come, the mobile homes at the camp site, I used to go and watch them'. Just like they have a day called 'the arrival-of-the-puffins day', they have got an arrival-of-the-tourists day. The tourist season here is short and intense, from June to August, and the tourists come and go like migrant birds. A younger woman talks about this as an important aspect of local life: 'Just see what happens when they open the café here (at the harbour) in summer – a lot of people come down here just to watch tourist life'. It is the same in Målselv – when the season starts at the ski resort the many visitors also attract local people: 'Both young and elderly people often come to the ski slopes just to watch the number of people'. The locals watch the tourists who watch them, as well as the puffins, the whales, the alpine mountains and the 'exotic' landscapes.

Even if the inhabitants don't interact much with the tourists, tourism seems to influence their perception of place. Local people don't only watch the tourists, they also look at what the tourists are viewing. Inhabitants in Andøy and Målselv say that they have become more conscious and proud of their natural surroundings: 'Stories are told by people visiting, about how fantastic a place this is, about the amazing nature'. Many told us how such positive stories by visitors became part of their own stories and perceptions of the landscape they live in and with: 'The stories are told again and again, and slowly I think they make people here proud, it influences the daily life of ordinary people'. The 'magic landscapes' also become their own.

Another important aspect is that the presence of the tourists makes these rural areas more vivid. In Målselv, for instance, the ski resort has created meeting places both between tourists and locals, and between locals: 'It's a new activity in the village. And a meeting place for many. There people meet fellow villagers. And can take a trip on the slope'. The cafeteria has become a social arena for inhabitants, where some come skiing and others just to have their Sunday dinner. This, many argue, has made Målselv more attractive, especially to the young: 'All of a sudden there is something cool happening in Målselv. Here you can go after-skiing, meet young people from the city every weekend'. It is claimed that 'we have become more attractive as a place of residence because there are more things happening; we've become more urban'. In Andøy, too, many emphasize the value of tourism creating more 'hustle and bustle' in the community: 'It makes this a better place to live, that we have more activities we can join'. And many local services, like local stores and cafés, depend on the tourist season: 'The sales in June, July and half of August are what we live on the rest of the year'. These are again crucial for many other encounters within these communities.

The strongest stories of encounters are those of small-scale tourist entrepreneurs. The direct meetings and personal relations they make with their customers are often their most important motivation for performing tourist activities. As one of them said: 'I want the tourists who come here to experience the place'. These personal meetings are also the strength of their enterprises. As the owner of a small campsite in Andøy states: 'The feedback from the tourists is that here they meet me'. Like many other small-scale tourism hosts he spends time talking to the visitors, and offers them some of his own knowledge and love of his village. The owner of a clothing shop explains that 'many tourists appreciate it; that it's a small fishing village where local people dare to be curious and engage with them'. These entrepreneurs are themselves an important part of the attraction offered. And the tourists become a part of their landscape.

When asked about conflicts, most people we talked to in Andøy and Målselv claim that tourism development involves conflicts to a minor extent. There seems to be a general opinion that 'tourism is good for the place'. But digging further into the relations involved in tourism development processes reveals many disputes and conflicting interests. Many local actors put forward the criticism of a lack of integration. In focus groups and meetings the lack of collaboration among tourism enterprises is often emphasized, and also the lack of collaboration between the tourism industry and other local enterprises, services and interests. While there is tight interaction between a few actors, others miss more cooperation. It is often stated that 'up here we have to cooperate to make people come' and that it would be 'a strength to all if we stand together', but it seems difficult to establish broad meeting places and platforms of collaboration. As will be elaborated in the next section, these contestations are related to complex discourses of community development (Saarinen 2004).

Discourses of Community Development

Continuities and ruptures: Pains and gains of the restructuring process

The process of restructuring involves the recreation of local geographies, and tourism development inevitably implies landscape changes (Wall 1996; Burns and Novelli 2006). As emphasis is put on new possibilities, established meanings, practices and relations are challenged. Both in Andøy and Målelv we were told many stories about the challenges of readjusting a stagnating military community. In both municipalities an object for the strategic community development work in recent years has been to redefine the municipality from military-dependent to hosts for the military, and to compensate for the loss of jobs. The political leaders stress that successful restructuring cannot be measured only by counting jobs: 'We measure as much by counting the mental absorption of change in the community, that we have to change, that we cannot remake what we had 30 years ago'. The restructuring processes are somewhat ambivalent. Many of the challenges these municipalities face are results of national and regional political achievements. The readjustment of the army and the structure rationalizing in farming and fishing are politically intended. 'The peace has come; it should be a positive thing'. But these processes hit hard in municipalities like Målselv and Andøy. A local politician explains that the negative population development was not hard to predict: 'It is a result of the hard work of establishing better infrastructure, roads, bridges, airports, which was seen as a formidable progress. But today we oppose the effect of this development. When it came, people started to use it, right? And we cannot wish ourselves back to a situation where bridges are blown up and the mastic ripped off the roads'. There seems to be a continuous tension between working to realize new ideas and to secure existing structures, which is expressed in various ways by politicians, municipal administrators and tourism entrepreneurs, as well as other local actors.

The introduction of new tourism activities has in some cases been seen as a threat to existing industries. In Andøya, when Whale Safari was established, the initiators were met with scepticism. Some describe it as a 'cultural clash' when Swedish researchers launched the idea of developing whale tourism: 'We were a fishing community, and it was like ... tourism could not just come and displace the fisheries'. This 'clash' was especially actualized when Whale Safari needed space to develop, leading to a heated debate about whether the harbour was and should be for fisheries only. This seems to have changed: 'Slowly there has been a mental awakening, that tourism actually can develop to become an important industry. Today it is not even debated'. The fisheries are not directly a part of the tourism development, as fish tourism is scarcely developed and strict rules prevent the fishermen from taking tourists on board. But most of the fishermen find the coexistence uncomplicated, as they have different seasons. Many also see tourism as a potential ally in their resistance to offshore oil drilling in their region. This is another highly contested topic of community development, and the fishermen find the municipality's strategy contradictory: 'They have given fisheries first priority, with tourism as a number two, but still they vote for researching possibilities for oil activity'.

In Målselv there has been a debate about whether the development of ski tourism is at the sacrifice of the grazing land for farmers and reindeer herders. As described by Nyseth and Viken (forthcoming), the conflict was to a large extent solved through a model of collaboration where reindeer herders were included as partners, along with landowners, with equal shares. They were also given the possibility of being part of the tourist attraction, offering reindeer sledging and Sami culture. Still, there are many challenging aspects of this development, which are not eliminated by economic benefit. In the documentary *Reindrift i kamp*[7] (Andersen and Ricardsen 2006) the delicate situation the reindeer herders were facing was portrayed. For years they had been fighting against the military's plan of expanding their training areas directly through their grazing land. Cooperating with the initiators of Målselv Fjellandsby, who 'had the power in relation to the municipality', became a key to solving this conflict. The reindeer herders' claim of the military reducing their expansion plans in order to accept the establishment of a ski resort was met, and a new agreement was set up. The herders felt they had no choice but to sign – the alternative was expropriation. Some of the reindeer herders expressed discomfort about having been 'bought and paid for' by the tourism industry, and are worried about the future for their form of life. Having signed the agreement with the ski resort, one of the herders reflected: 'Did we really do the right thing, thinking of our descendants?' The disputes with traditional industries have to do with identities as well as the rights to natural resources.

Growth and balance: Diverging development objectives

The focus groups and discussions in various meetings about tourism development revealed quite divergent and sometimes conflicting aims regarding tourism development. As argued by Saarinen (1998; 2004), discourses of development are

---

7   Literal translation: Reindeer herding in struggle.

many. Whereas the larger enterprises are preoccupied with increasing the volume, many of the small-scale enterprises are concerned with avoiding the place being 'too touristy'. Målselv Fjellandsby aims at getting charter flights: 'It's charter flights we have to go for. The structure of Fjellandsbyen is based on volume'. As a first step, the ski resort concentrated on the regional market, selling vacation-home lots and ski passes to people from the region. Later they wanted to increase their market and attract foreign tourists. The managers of the ski resort see charter flights as the only way to get the required volume: 'Fly in and fly out, that's how we can make a profit'. In Andøy, Whale Safari also sees charter flights as an objective, to exploit the capacity on the boats better and to expand the season. They have plans for increasing their capacity, to create a 'hall of giants', with full-sized whales, and maybe to build hotel facilities. In recent years they have, in cooperation with the municipality and other actors, worked directly towards airlines establishing direct flights to Andøy, and see a possibility for safari tourists 'flying in in the morning and out in the evening'.

This strategy is highly contested by others. Many local actors in Andøy and Målselv express resistance to 'developing mass tourism'. Andøy is part of the region of Vesterålen, which is experiencing strong growth in tourism. Vesterålen is often called 'the new Lofoten'. In Andøy, many question this development; they 'do not want to become touristic like Lofoten'. There are many stories of people, both tourists and inhabitants, who seek Vesterålen out to escape the mass of tourists in the neighbouring region. They talk about Lofoten as a place that 'has become industrialized', where 'they have started to build upwards to store away people' and where 'everything is commercialized'. It is argued that 'they use up the place (...) it is not what we should aspire to'. Even if most people seem to agree that Vesterålen 'has much unused potential', they are concerned with maintaining their quality and distinctive characteristics. A small-scale tourist entrepreneur states: 'This place has a soul, really, there are so many soulless places, but here it is not so touristic'. But this valued landscape is conceived as vulnerable, as the tourist industry develops and many houses are turned into vacation homes. Many emphasize the importance of 'remaining a vital fishing community'. In a meeting where local tourist operators were discussing development potential, several small-scale entrepreneurs expressed concerns about fast growth at the expense of these qualities: 'There has to be a balance'.

In Målselv there are parallel discourses. The development of the ski resort and the destination Snowman Area is very much inspired by the ski destination of Levi, which is the biggest ski resort in Finland, and by the marketing of the Santa Claus village in Rovaniemi: 'Everything we do, we copy from Levi' (see also Viken in Chapter 7 of this volume). Others are worried about this development: 'They look to Levi, what they have managed to create there. But I would say God help us not to become like Levi, because that is just so ... '. The most critical voices come from small-scale tourist entrepreneurs and environmentalists, who oppose the strategy of charter traffic to Målselv. They claim such a development is incompatible with the exotic and pure destination they are offering. They also question the environmental aspects of this development, claiming that charter tourism will contribute to climate changes and melting snow: 'They go for the snowman, but for every tourist they bring in to look at it, the snowman will melt a bit'. Many small-scale operators feel that the focus on tourism growth has made the municipality, the public support system and regional destination companies 'lose sight of small-scale actors'. Those promoting this development tend to explain the resistance as typical for dalesmen[8], described as 'arch conservative' and 'resisting anything new'. In the dominant discourse of development, the conditions for critical voices, promoting alternative landscapes, seem difficult.

Tools of development: Innovation and anchoring

Another contested aspect of tourism and community development is the role of public administration and policy. As demonstrated by Bærenholdt et al. (2004), local authorities often play many and crucial roles in tourism development. In Målselv the active role of the municipality, and especially of the mayor, is seen as a key to the development of the ski resort and many new enterprises. The municipality of Målselv has spent substantial resources, both administrative and economic, on the establishment of Målselv Fjellandsby. Many emphasize the many creative and flexible solutions that were made. The mayor explains: 'We have constantly worked in unconventional ways to establish new activities'. The initiator and main investor of the ski resort required support from the municipality in order to build in Målselv. He got a promise from the mayor that: 'If you need a political resolution, even in 15 minutes and on a Sunday, just call'. In this way, 'the political and bureaucratic barriers were cleared away'. The regulations required were made at high speed; in Målselv they talk of a national record. In 2007 Målselv was elected Norway's most innovative municipality for this work. But the municipality's use of administrative and economic resources is controversial, and so are the unconventional proceedings. In order to 'get things done', the municipality has been 'tempted to take some shortcuts'. The high speed has a cost: 'Along the way we've been criticized for the speed, that it became undemocratic. It was not like we sent a letter, we ran along, travelled to Oslo with the letter'. The establishment of the ski resort became the main issue for local development, and the municipality is accused of giving priority to Målselv Fjellandsby at the expense of other development projects: 'They have in a way just jumped the queue in many fields. And that creates dissatisfaction'. Many local interests, private persons and people within other industries and the voluntary sector, feel neglected.

In Andøy the municipality has played a less significant role in the tourism development. Here it is debated whether tourism actually can be seen as an area of priority. Several of the tourism operators claim it is not. The mayor explains that although they have agreed on a strategy, they lack tools for acting on it. And that it is difficult in a fishing community to gain support for reallocating resources to new

---

8   In Norwegian 'døler'.

activities: 'If we are to give priority and contribute, the distribution has to change; less to fisheries and more to tourism and other industries'. While they have special loans and financial support for fishermen, they lack such incentives for other industries. All they can offer is help related to the public support system and a small amount of money. And the public support system, like Innovation Norway, is again criticized for lack of flexibility. While many of the tourist enterprises have got substantial economic support, others talk about strict rules making it difficult for them to obtain funding.

An important tool of tourism development in Andøy is the newly opened National Tourist Road. This garners a lot of publicity, and also considerable investment by the National Road administration. There are high expectations in the municipality of what this can lead to, but there is also resignation. Local people tell how they were invited to a consultation process, where mass meetings were arranged and the inhabitants delivered their suggestions of what should be done. One of the local tourist operators explains: 'We participated and engaged, and came up with suggestions. But then we heard nothing. We were neither connected nor disconnected'. Many have become frustrated, as they experience a process which was 'far from as democratic as it was first presented to people'. The planned installations, architect-built vantage point, rest stop and public house, are somewhat diverging from the inhabitants' wishes. As also demonstrated by Saxena et al. (2007), it seems like lack of integration and local anchoring creates tensions over tourism innovations.

Place branding: Profiles challenging local identities

Tourism development involves creating new representations of place. As has been emphasized, places have multiple identities, and representations of place are always partly and thus highly political (Massey 2005; Saraniemi and Kylänen 2011). Studying tourism development in Andøy and Målselv reveals many representations and brands. In Andøy, they have an official brand, *Nordlyskommunen* (the municipality of Northern lights), which has not been much used. At the municipality's tourism web-site, Andøya is profiled as 'the magic island'. But this is not a brand used by the tourism enterprises or others in the community. Rather, we found that few had heard of it. Whale Safari use Moby Dick as their main brand. Other enterprises on the island use descriptions such as 'coastal wilderness', 'grand nature' and 'magic adventures'. In Målselv, the official slogan of the municipality is 'The land of possibilities'. In the tourism industry many descriptions are used, like 'Arctic Lapland', 'winter paradise', 'Arctic adventures' and 'Sami adventures'. Connected to the new ski resort, there have been many brands during its few years of existence; *Myreffell, Blånisseland, Målselv Fjellandsby* and the *Snowman area*.

These conceptualizations used by the tourism industry do not mirror the inhabitants' perception of place. Whereas representatives from destination companies and leading figures of tourism development argue for more coherent branding, and recommend that the tourism industry agrees on 'the most effective brand', local people insist on multiple local identities. In Andøya most people characterize the community as 'a fishing community', but stress the variety in nature and social life. Many emphasize 'the rough life by the great ocean', others 'the peace and quiet' of a rural society. The entrepreneurial activity connected to space research and youth culture is often referred to. The aspects emphasized differ between actors. It is the same in Målselv. When asked to describe Målselv, most local people emphasize 'the spectacular nature, the beautiful mountains, the woods and the river'. Many also underline 'the vigorous cultural life, with solid traditions of folk music and food' and 'the solid sport milieu and traditions of voluntary community work'. Some point to 'the creative milieus' for filmmaking and snow ice sculpting that has developed in recent years. The characteristic 'dalesmen culture' is often used, referring to the ancestors migrating from the valleys in South-East Norway. The dalesmen are seen as 'proud' and 'reliable', but also 'slow' and 'conservative'. At the same time Målselv is perceived as a military community with 'an intense mixture of people and dialects' and high mobility and flexibility. There is thus a great diversity of local identities.

The effort to create a destination brand, like the Snowman area in Målselv, enters into this complexity. Although celebrated by many, the brand was also met with strong resistance by a range of local actors (Førde, forthcoming). Elderly people accused the branding of having 'no sense of history' and young people found it 'kitschy'. They did not want to live in 'Snowman land', and felt their local identity to be threatened. Many local actors also felt excluded in the process of developing the brand. Representatives of local sports associations, cultural associations and the snow and ice sculpting milieu claim that their competences have been ignored, and that the Snowman area thus becomes a brand that 'does not at all signify where we are in the world'[9]. Another aspect is the continuous tension between different localities in the municipality. Both Målselv and Andøy consist of numerous villages, which have different characteristics and are involved in continuous struggles: 'Målselv is not *one* place ... The east and the west side of the river do not necessarily have congruent interests'.

Yet another aspect in these negotiations of representations is related to class. Both in Målselv and Andøy the discourses of tourism development encompass a continuous debate about quality and standards. These discourses reveal power structures of symbolic capital. In Målselv the new development in Målselv Fjellandsby, with high-standard cottages, is often contrasted to the more unassuming establishments down by the cascade. Here they offer basic, lower standard accommodation, arrange dance festivals, singles festivals and the like. Representatives of the new tourist activities don't want to be identified with these popular services and events that they find 'simple', 'cheesy' and 'without style'. The contrast became evident when local actors wanted to set up caravans in Målselv Fjellandsby: 'They did not want caravans there'. The result of this negotiation was

---

9   As pointed out by Viken in this volume, the Snowman label was later abandoned.

that they found a place for the caravan tourists at the bottom of the ski slopes. In Andøya the recent opening of the National Tourist Road raised a debate about the quality of local food and accommodation services. In national media as well as in local public arenas it was claimed that the standard of these were 'too simple', and that the island does not have 'proper restaurants'. This is perceived as an insult by several of the local enterprises. As Picard (2011) demonstrates tourism development initiatives demand cultivation of local social life and triggers new forms of self-understanding.

Conclusions: Negotiated Landscapes

Tourism represents a promising alternative to the rural decline in Andøy and Målselv. But it also entails transformed and negotiated landscapes. Applying a relational approach to tourism development, I have in this chapter tried to explore how these geographic landscapes are reproduced through a range of networking practices, as material natures, social relations and discursive conceptions. Focusing on the interrelation between the tourist landscape and the inhabited landscape of local experience, through encounters and discourses of community development, I have examined how tourism development enters into a complexity of on-going local processes.

Tourism and community development causes contestations over landscapes. As Ringer (1998) argues, the perception among local people that they live in a distinctive landscape is essential to their creation of community and a well-defined sense of place – of sharing selected mental and physical characteristics of that particular locality. The perceived landscapes constitute the framework for establishing local ways of doing and seeing. Transforming a whale-hunting boat into a boat for whale safaris, a fishing industry plant into a restaurant and a stockfish-storage into a dance bar are thus actions of landscape politics. So is transforming sheep and reindeer pastures into ski slopes. The transformation of symbolic, social and material landscapes in order to generate tourism implies processes of inclusion and exclusion, and of local identities being challenged.

In 'the magic island' of Andøya and 'the exotic alpine destination' of Målselv tourist landscape and the inhabited landscapes are meshing, but also clashing. The creation of destinations enters into broader processes of community development and causes negotiations of local landscapes. If tourism is to be a viable contribution to these communities, the complex relations between tourism landscapes and inhabited landscapes have to be taken into consideration. The exotic and the ordinary are indissolubly interwoven.

References

Bærenholdt, J.O., Haldrup, M., Larsen, J. and Urry, J. (2004), *Performing Tourist Places* (Aldershot: Ashgate).

Berg, N.G., Dale, B., Førde, A. and Kramvig, B. (2013), 'Introduksjon: Metodologiske utfordringer i stedsanalyser', in Førde et al. (eds).

Burns, P. and Novelli, M. (2006), 'Tourism and Social Identities: Introduction', in Burns and Novelli (eds).

Burns, P. and Novelli, M. (eds) (2006), *Tourism and Social Identities: Global Frameworks and Local Realities* (Oxford: Elsevier).

Butler, R. (1999), 'Problems and Issues of Integrated Tourism Development', in Pearce and Butler (eds).

Frisk, L. (1999), 'Separate Worlds: Attitudes and Values Towards Tourism Development and Co-operation Among Public Organisations and Private Enterprises in Northern Sweden', in *Proceedings of Forskarforum. Local och regional utveckling conference*, Öresund, 16–17. November.

Førde, A. (2011), 'From Fishing Industry to 'Fish Porn'. Tourism Transforming Place', in Macleod and Gillesppie (eds).

Førde, A. (Forthcoming), 'Entrepreneurship and Controversies in Tourism Development' in Johannesson et al. (eds).

Førde, A. et al. (eds) (2013) Å *finne sted. Metodologiske perspektiver i stedsanalyser* (Oslo: Akademika forlag).

Hughes, G. (1998), 'Tourism and the Semiological Realization of Space', in Ringer (ed.).

Ingold, T. (1994), *Companion Encyclopedia of Anthropology: Humanity, Culture and Social Life* (London: Routledge).

Jóhannesson, G.T. et al. (eds) (forthcoming) *Tourism Encounters: Ontological Politics of Tourism Development.*

Jóhannesson, G.T., van der Duim, R. and Ren, C. (2012), 'Introduction', in van der Duim et al. (eds).

Kramvig, B. (2007), 'Flerstedlig og flerstemt – som situeringsforsøk i lokalsamfunnsstudier', in Nyseth et al. (eds).

Law, J. (2004), *After Method. Mess in Social Science Research* (Oxon: Routledge).

MacLellan, R. (2011), 'The Role of Destination Management Organizations: Models for Sustainable Rural Destinations in Scotland?', in Macleod and Gillespie (eds).

MacLeod, D. (2011), 'A Canary Island in the Context of Sustainable Tourism Development', in Macleod and Gillespie (eds).

MacLeod, D. and Gillespie, S. (eds) (2011), *Sustainable Tourism in Rural Europe* (Oxon: Routledge).

Massey, D. (2005), *For Space* (London: Sage).

McCabe, S. and Marson, D. (2006), 'Tourists' Constructions and Consumption of Space: Place, Modernity and Meaning', in Burns and Novelli (eds).

Nyseth, T. et al. (eds) (2007), *I disiplinenes grenseland. Tverrfaglighet i teori og praksis* (Oslo: Fagbokforlaget).

Nyseth, T and Viken, A. (Forthcoming), '"Bon-fire Diplomacy"? Community Conflict Solving through a Destination Development Process'.

Pearce, D.G. and Butler, R. (eds) (1999), *Contemporary Issues in Tourism Development* (London: Routledge).

Picard, D. (2011), *Tourism, Magic and Modernity* (Oxford: Berghahn Books).

Ringer, G. (ed.) (1998), *Destinations: Cultural Landscapes of Tourism* (London: Routledge).

Ringer, G. (1998), 'Introduction', in Ringer (ed.).

Robinson, M. and Smith, M.K. (2006), 'Politics, Power and Play: The Shifting Contexts of Cultural Tourism', in Smith and Robinson (eds).

Saarinen, J. (2004), '"Destinations in Change": The Transformation Process of Tourist Destinations', *Tourist Studies* 4:2, 161–179.

Saarinen, J. (1998), 'The Social Construction of Tourist Destinations: The Process of Transformation of the Saariselkä Tourism Region in Finish Lapland', in Ringer (ed.).

Saraniemi, S. and Kylänen, M. (2011), 'Problematizing the Concept of Tourism Destination: An Analysis of Different Approaches', *Journal of Travel Research* 50:2, 133–143.

Saxena, G., Clark, G., Oliver, T. and Ilbery, B. (2007), 'Conceptualizing Integrated Rural Tourism', *Tourism Geographies* 9:4, 347–370.

Smith, M.K and Robinson M. (eds) (2006), *Cultural Tourism in a Changing World: Politics, Participation and (Re)Presentation* (Cleveland: Channel View Publications).

Tinsley, R. and Lynch, P.A. (2008), 'Differentiation and Tourism Destination Development: Small Business Success in a Close-Knit Community', *Tourism and Hospitality Research* 8:3, 161–177.

van der Duim, R. et al. (eds) (2012), *Actor-Network Theory and Tourism. Ordering, Materiality and Multiplicity* (Oxon: Routledge).

van der Duim, R., Ren, C. and Jóhannesson, G.T. (2012), 'Tourismscapes, Entrepreneurs and Sustainability. Enacting ANT in Tourism Studies', in van der Duim et al. (eds).

Viken, A. (2013), 'Fokus på sted – fokusgruppesamtale som metode i steds- og destinasjonsanalyser', in Førde et al. (eds).

Wall, G. (1996), 'Integrating Integrated Resorts', *Annals of Tourism Research* 23:3, 713–717.

Chapter 10

Standardization and Power in Cruise Destination Development

Ola Sletvold

Introduction

How does getting involved with cruise tourism impact on land-based production and cooperation in a particular destination? This chapter is intended to shed some light on cruise excursion production in Lofoten in Northern Norway. It describes how land-based tourism industry actors and other community members can present and reflect upon their responses and adaptations to the cruise industry and tourists and on the development of Lofoten as a cruise destination. Although it is well established, this destination is a sum of small, spread out and vulnerable businesses and communities. The increase of cruise tourism may imply standardization in the sense coined by Henry Ford – 'any color as long as it's black' – (Ford 1922), meaning less variation and more uniformity, bigger volumes and lower prices. The chapter will present the production characteristics of Fordism and McDonaldism as applied to cruise tourism and land-based cruise excursions and discuss their impact on the destination of external standards that follow cruise tourism.

Change in the sense of adaptation or adjustment implies that there will be bargaining concerning prices, volumes and time. Business relations where local and global actors meet are often presented as asymmetric power relations between small local companies and large corporations (Ford et al. 2011; Hall 1994) resulting in transferring the will of cruise companies to compliant local hosts. Others claim a more changing or floating picture of power at work (Milne 1998). The latter position is related to the Foucauldian understanding of power as fluid and multiple, located in networks of relations and repressive and productive at the same time (Foucault 1977; Cheong and Miller 2000). This paper seeks to trace the picture of power expressed through producers' recounts of the development process.

According to Klein (2012) much of the research relating to cruise tourism is technical and practical in nature and tends to 'support the interests ... of the cruise industry', as opposed to 'emancipatory research' (Klein 2012, 7). The present text aims to take a different and more empirical route, by listening to local businesses, restaurants, museums and attraction producers, transporters and coordinators, in addition to community members. The chapter first includes some background on the destination area Lofoten. Some basics of the cruise industry and its structural features of production are presented. Some ways in which power can be exerted and created in cruise tourism are also examined. The methodological considerations underlying the research are presented and this leads into discussions about the ways in which the destination actors have been influenced by increased cruise tourism.

The Lofoten Destination Area

The Lofoten archipelago at 68° north stretches into the Atlantic from mainland North Norway, with a rugged Alpine topography of mountains rising from the sea to above 1000 meters, interspersed with some areas of agricultural land. The mild climate is due to the Gulf Stream and easy access to fish resources has formed the basis for settlements and traditional ways of living for more than 1000 years. In particular, the seasonal cod migration for spawning has attracted fishermen from all over Northern Norway. Fish is hung outdoors to dry for export to Southern Europe, a trade that for centuries has linked the Norwegian coast to the trade systems and cultures of continental Europe – and still does. Since before its Viking past (AD 800–1000) Lofoten has been an important farming region.

Fifty years ago Lofoten could only be accessed by boat. Today you can drive across the five main islands via bridges and tunnels, and there are airports and express boats (cf. Figure 10.1). One main road takes visitors 140 km through the region, between the biggest town Svolvær in the east to the fishing village Å in the west. Minor roads link villages to the main road. There has been a regional centralization of the population (today 23,000) towards municipal centres and the service industries have become a dominant sector. During the last couple of decades Lofoten has become a well-established tourist destination. Land based tourism is mainly a summer season and round trip phenomenon. There were approximately 60,000 bed nights in 2005, while the direct accommodation income was estimated at 430 mill NOK, tourist consumption at 630 mill NOK and the total effect at 800 mill NOK (Destination Lofoten 2006, Tables 1.3 and 1.5).

Lofoten's prominent attraction is the magnificent landscape with its quaint fishing village settlements – the iconic *rorbu* housing and drying-racks for cod – supplemented by planned attractions like a major Viking Age and other minor museums, several art galleries and handicraft producers, and fishing, skiing, climbing and hiking activities. *Rorbu*s converted for accommodation and restaurants established in former sea warehouses are a speciality.

Lofoten scarcely figures at all in international cruise statistics. The sophisticated marketing strategies of the industry, such as 'supersegmentation' (Ioannides and Debbage 1998, 112) and Internet based market penetration (Weaver 2008) characterize the companies' targeting of their market communication. Cruises visiting Lofoten are included in the North-West Europe category – the Baltics and Scandinavia including Svalbard (Charlier and McCalla 2006), which represents around 6–7% of the global market. The amount of cruise traffic arriving in Norway was 1.4 million harbour visitors (430,000 unique visitors) in 2009, rising to 2.5 million (588,000 unique visitors) in 2012. The number of ship calls rose from 1,506

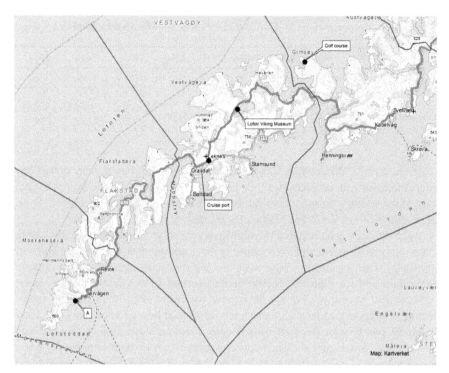

Figure 10.1   Map of Lofoten

Source: Statens Kartverk

(2005) to 2,066 (2012) (Farstad and Dybedal 2012; Cruise Norway 2013). The cold water aura of North Norway makes cruises somewhat specialized, mostly using middle sized and smaller vessels. An account of the Lofoten traffic for the last nine years is presented in Table 10.1. During the ten years before 2004 the number of ship calls varied between 38 and 70 and the forecast for 2013 is 86. Doubling the visitor numbers in a decade reflects the general development towards bigger ships and may represent an important push in the development of an already well-established destination. In addition to passengers there are the crew visitors, but there are no statistics for these. The normal ratio on board is one crew member per two–three passengers.

Cruise tourists are volumes added to land based traffic and tourist accumulation at certain times and spots quickly creates overcrowding. Fishing villages surrounded by the ocean, with only a limited number of roads, attractions and facility capacities like toilets and parking space may have a problem: At Leknes port, the 2012 season saw 48 days of cruise visits during the short summer, seven of these days with more than 2000 passengers. On the most crowded day they planned to serve 3,600 visitors. Such land side cruise production would be chaos without careful destination logistics. Leknes is centrally placed – it takes one hour by bus to the western end of the road in Å, and the same to Svolvær in the east

Table 10.1   Cruise visits to Lofoten (L=Leknes, S=Svolvær) 2004–12

	2004	2005	2006	2007	2008	2009	2010	2011	2012
Ship calls	70	72	53	77	85	73	L 54 S 14	L 38 S 18	L 65 S 21
Passengers	34000	35700	34590	35360	55387	52284	L 46333 S 5743	L 50389 S 6893	L 60737 S 7854

Source: www.cruisenorway.no

(cf. Figure 10.1). On average, 76% of all cruise passengers buy full day or half day shore excursions so on such days they need more than 60 buses and guides to take care of the visitors, far more than can be solved with the regional resources. Thus, producing Lofoten for cruise visitors faces challenges of mass tourism.

Infrastructure is a recurring theme whenever cruise development is discussed. Ports may wish to expand their capacity to accommodate ever bigger ships at quaysides. This is a matter of efficiency and logistics with regard to access to shore excursions and for supplies. The optimum measure from a cruise industry point of view is to have special quays designated for cruise ships, with an ISPS fence, technical support for electricity and fresh water supplies, garbage and grey water handling, comfortable transfers for passengers and crew and facilities for queuing and waiting, toilets and shelter, internet access etc. A harbour is normally not a cruise terminal. Of the approximately 40 harbours in Lofoten, four have registered cruise quays, and Leknes port is visited by 80% of all cruises. In 1984 the manager of European Cruise Service (ECS) visited Lofoten and found the Buksnes Fjord outside Leknes (cf. Figure 10.1) to be an excellent harbour. It is shielded from most winds, is deep and wide enough for easy manoeuvring and has good anchoring conditions, which makes it a logistically ideal place to start excursions. In the 1980s they still had to carry out transfers using tender boats. In the early 1990s the wishes of the cruise industry for a quay at Leknes that can accommodate cruise ships were met. The quay is for general industry purposes, and there is a fish products factory, stock houses for potatoes, transport facilities, and 'Oddny's Little Tourist Shop'. This was a crucial step in infrastructural development.

The fact that cruise ships call at several ports in Lofoten illustrates the municipal authorities' willingness to engage in this industry (and their interest in the port tax income), so that they can apply for the government financing of quay development. For a small municipality, the Leknes port tax is considerable at some 600,000 NOK. At the same time the spread of cruise calls reflects the still decentralized societal structure of Lofoten and the accompanying spatial structure of the tourism networks in the destination. The region is definitely a periphery, but the traditional fishing village landscape is alive even off seasons, with small size fish production, local shops, commuting and school bussing.

Itineraries along the Norwegian coast and in the fjords are motivated by sightseeing of the landscapes and the inhabited areas. The quality of the experiences depends largely on land-based inputs in the cruise and Lofoten is seen as Norway's third biggest attraction for cruises, following the Fjords and the North Cape. These rural areas are very different from city destinations with regard to infrastructural capacities and the number and types of businesses required for handling cruise traffic, resulting in infrastructural stress (Chin 2008). A city landscape can take thousands of visitors without compromising its character.

Cruise Industry Characteristics

Cruise tourism is a very globalized phenomenon (Wood 2000). It is literally a flow of technology, people, information, images and money, resulting in experiences produced on or near by water almost anywhere on the globe. It has been growing rapidly for the last 30–40 years (Hobson 1993; Dickinson and Vladimir 1997; Weaver 2005). The annual growth in passengers has averaged 8% since 1980 (Miller and Washington 2010, 138), rising from 9.5 million to 14 million at the end of the century (Douglas and Douglas 2001; Wood 2004). A market forecast for 2013 is 20.9 million (Cruisemarketwatch 2013). This growth is due to capacity increases and ever bigger ships being built, along with considerable dynamics in ship design innovation and technology development (Jordan 2008), as well as market expansions, mass market pricing and marketing segmentation (Weaver 2011), related to itinerary and excursion innovations (Dowling 2006) and theming (Miller and Washington 2010; Weaver 2011). The industry is described as an oligopoly (Wie 2005; Bezic and Vojvodic 2010). The three leading consolidated corporations (Weaver and Duval 2008) have an 85–90% market dominance (Kwortnik 2006; Miller and Washington 2010,139). The biggest company, Carnival Corporation, had a portfolio of 11 brands and carried nine million passengers in 2010 (Datamonitor 2011) and has become a leading innovator in the industry (Kwortnik 2006).

The literature on the cruise industry identifies some characteristics that are important for a picture of this industry in the interaction with the land-based tourism industry. Ionannides and Debbage (1997; 1998) initially claimed the cruise industry to be 'yet another Fordist representation of the travel industry', characterized by horizontal integration and economies of scale. Fordism is Weberian rationalization conducted in industrial and service economies (cf. Ritzer 2002) and its often mentioned traits include: product standardization, inflexibility/rigidity, mass replication, small numbers of dominant producers and mass marketing to an undifferentiated market (Torres 2002). Contemporary tourism, however, has seemed to develop towards a rather complex situation with great product differentiation, flexible specialization, mass customization and fragmented markets. Torres (2002) and Ioannides and Debbage (1998, 117) have concluded that the travel industry is 'neo-Fordist', implying that Fordist principles coexist with post-Fordist ones. Along the same lines, Chin (2008, 12) claims that niche markets are nested within mass markets.

Weaver (2005) has shown that production on board mega cruise ships has several signs of McDonaldization (Ritzer 2002): efficiency, calculability, predictability and control, while at the same time demonstrating the Fordist principles of being large-scale, homogenized and standardized (Torres 2002). But there is also the post-Fordist trait of customization through the enormous number of options available (Weaver 2005, 349). That cruises carry the main characteristics of Fordism and McDonaldization needs further nuances. Cruise markets are highly competitive and fragmented through operators who cater to any specialized demand, notably within the expedition, luxury and niche markets. The volume of passengers needed to make such a cruise feasible is therefore far from the 6000 needed to fill the largest ships. Thus, one could expect to see post-Fordist flexible specialization and individual customization alongside neo-Fordist mass customization (Torres 2002; Weaver 2005).

Land excursions have become important in the cruise competition. They involve profit potential, itinerary differentiation and some customization. Excursion manuals produced in cooperation between the destination partners present trips being available only through booking on board, and the success of excursion sales heavily influences the total profit of a cruise. The calculability character of this operation is also demonstrated by the prices tourists pay, an often mentioned rule of thumb is twice the price agreed between cruise agents and land-based producers. Together with the logistics operation of excursions, supplying cruise ships with all the goods needed once they have left the port adds to management challenges in planning going beyond those of a hotel (Véronneau and Roy 2009).

The question that arises from this general picture is how the above-mentioned characteristics of cruise production manifest themselves in a specific destination in Arctic Norway. How do they impact on business relations between land-based producers and the cruise lines, and on the cooperation within the destination and on views on cruise related development? To ask for impacts implicitly is to look for influence and to say that processes have aspects of power. Power in tourism development has been studied from a strategic planning and cooperation perspective (Reed 1997), following the Lukes (2005) tradition of looking at agenda control and conflict of interests, and as asymmetries within networks that exchange critical resources (Ford et al. 2011). These perspectives have a bearing on the present case, in which the attention given to cruise related matters in strategic destination planning is questioned, and where the importance of information resources is acknowledged with regard to the adaptation to cruise expectancies. On the other hand, Cheong and Miller (2000) make a general case for the fruitfulness of applying Foucault's conceptualization of productive power to the theme of tourism, as Marzano and Scott (2009) have done with Foucault in a study of destination branding.

If power is omnipresent (Foucault 1980) relations on board a cruise ship are also relations between 'targets and agents' (Cheong and Miller 2000); the dominated and the domineering. The tourist targets are free to wander around and to choose and pick but the carefully programmed days of cruises define quite limited scopes of freedom. The predictability, efficiency and mass replication of cruise trips are aimed at maximizing passengers' consumption. The on board crowding and queuing involves disciplining of the tourist targets, although the repertoire of entertainment inside and the vistas outside the ship may be wide enough to prevent widespread boredom (Weaver 2008).

For our purposes, the more interesting agents and targets are the representatives of the cruise industry – agents and operators – and the land based producers. The negotiations that take place when the operators meet land based businesses aim at preventing the occurrence of later complaints through the presentation of specified orders for how, when, for how long and with what their passengers expect to be entertained. To follow Foucault (1980), that knowledge induces the effects of power: with reference to the argument about knowing their customers, their preferences, their standard expectations and willingness-to-pay representatives can present their demands. If standardization works through such mechanisms, we would also expect that the power exerted through the process enables power to be in the hand of the targets, which are the local businesses.

Cheong and Miller (2000, 379) also point out the role of middle-men, the brokers. Bus guides represent the cruise companies and present the excursion stops to the tourists, thus creating relations both ways during the production of experiences. Likewise, during the planning of excursions, there is a local cruise agent, who has relations with local businesses while conveying the market information that the industry wants to disseminate, regarding trends and innovations that might affect traffic to the destination. During cruise fairs, by arranging meetings, carrying out presentations, translating, handling questions or complaints or solving practical problems, brokers will affect the enabling of power in the processes.

To what extent then, can the people interviewed in Lofoten explain the impact of standardization and the power mechanisms embedded in their dealings with cruise operations? Before answers can be given, the people in question need to be introduced.

Methodology: Mapping the Cruise Destination Voices

A destination is the cumulative, produced result of marketing, information and image shaping processes. It is a sum of enactments and performances by tourists (Bærenholdt et al. 2005) and service personnel (Weaver 2005). An established tourist destination is very much the result of market oriented cooperative and coordinated processes within industrial, social and political networks in a geographical area (Sheehan and Brent Ritchie 2005; Scott et al. 2008; Lemmetyinen 2009) and of the accompanying continued reflections on and negotiations of place identity by people there (Saarinen 2004; Hultman and Hall 2011).

With the aim of getting a grip on development that has taken three decades, and that to a very small extent is told in documents, it was found reasonable to work along qualitative principles. Several understandings of the destination may be present, and the interpersonal dynamic and cross-referencing which is a potential in groups was preferred, for the purposes of time saving. The present case study is based on a series of semi structured group interviews and on secondary documentary information. A purposeful convenience sampling (Patton 2002) approach was used in the listing, grouping and recruitment of interviewees. To address the varying salience and centrality for the activity in focus (Sheehan and Brent Ritchie 2005; Scott et al. 2008) the main criteria used were geography (with the cruise port at Leknes as a node) and business relevance (excursion products providers and coordinators). Furthermore, municipal involvement in cruise development (political and administrative) and ordinary opinionated people were searched for. Official industry registers, cruise manuals, web sites and documents covering cruise activities were combined with previous knowledge of the region. This resulted in an extensive list of potential informants, and the task of recruiting interviewees was given to a local consultant. Eight groups were formed. There was some no-showing due to extremely bad weather, but 43 people participated. Nine group interviews were conducted over four days in early December 2011, resulting in 15 hours of recorded discussion.

The topics planned for the interviews were the history and success of cruise development in Lofoten; experiences in working with the cruise industry; business and public infrastructure for delivery of the cruise product and formalities and communications in cruise related co-operative constellations.

The interviews were transcribed, resulting in 300 pages of text. Repeated listening to the interviews was combined with reading the transcripts verbatim. It appeared that the group talks partly had the character of a discussion on strategy and on Lofoten's place and destination identity. The analysis was limited to the twin themes of single business adaptations to cruise demands and inter-business and destination level work with cruise tourism. The five groups with different business providers thus proved most fruitful, whereas the other groups became secondary, providing mainly supporting information.

Excursions Involvement

The market success and profitability of the cruise industry results from the systematic production principles at work, and this success has represented a motivation for the continued involvement of the land based providers. Specifically, cruise guests imply predictability. Now several say, 'we depend on cruises to survive' and excursionist guests are 'predictable in the budget'. A Lofoten restaurateur will for example know almost the exact number of lunch cruise guests at a precise time slot on a specific day one and a half years ahead and to serve 200–300 people in an hour is possible only with the same planning rationality and McDonaldist efficiency that cruise kitchens operate (Weaver 2005, 351). The meal is a standardized buffet (it is never a la carte), all costs can be calculated, the necessary personnel recruited, and the amount of waste is foreseeable, as are profits. This calculability of income and cost provides operational freedom, a basis for living with other seasonal uncertainties, 'an economic sole', some say, for weeks of low profit making. However, they still wish to see the single tourist: 'let them feel welcomed as individual people'. The money from volumes is attractive, but they try to relate to each customer: 'cruise passengers are individuals too', and 'if the operator tells of older or handicapped tourists, we seat them on the bottom floor and send the more mobile ones upstairs'.

The rational calculation and logistics control is illustrated by documents. The list of the season's cruise calls shows port related details of size, length, running depth, hours at the quayside or by anchor, and with tourist related information on numbers of passengers and languages spoken. Also, the schedule for the shore excursions list bus numbers, languages represented, driving distances and times to the stopping spots and the time allocated there. The planning even shows that different language groups are not expected to spend the same amount of time at the same place: Germans who do not eat spend one hour in Nusfjord, whereas fellow passengers from France who arrive at the same time spend half an hour. Whether this is due to bargained prices, a result of cultural differences, or both, it may be a case of flexible specialization – but it is certainly well planned.

Cruise standards also increase the efficiency of accountancies procedures; one single invoice serves many guests, and, as one put it, 'you only deal with the operator in Bergen'. However, some uncertainty is often expressed regarding economic advantages, due to the clause about last minute reductions in guest numbers without a parallel possible reduction in costs. Such experiences with cruise operations become part of next year's negotiation competence and blend with other mixed feelings. For example, a successful restaurateur tells about how he was first 'exploited' by cruise operators with 'zero focus on quality', chasing for the 'lowest price possible and biggest volume possible', implying messy cruise buffets with 'chaos' and 'tables (that) must be totally renovated' between sets of guests. Later, he chose to cancel his cruise involvement in one of his restaurants, as 'they wanted to dictate the food I should prepare' and stated that 'I represent a certain quality'. At another restaurant in another village, however, he still serves cruise tourists, as the particular village depends on this restaurant to sustain cruise visitors to other local attractions. He has strong opinions on what strategies and attitudes are needed in negotiations with cruise operators. He thinks that local actors need to realize their strength. Thus, for him, cruise involvement has induced an understanding of power.

Meeting cruise standards may lead to material changes as serving masses implies making access possible to new customer groups. At Lofotr Viking Museum amendments were made to their access area to make sure that quite elderly and less mobile cruise passengers can board and row the Viking ship. The good terms established show in negotiations with the operators who now begin by saying 'How many can you take between eight o'clock and five in the afternoon?' and plans for future renewals of their exhibitions, activities and events take cruise volumes into consideration. In traditional houses which were converted for restaurants or museum purposes, small rooms and steep staircases cannot be replaced or a lift installed that would make all areas accessible to everyone. Issues about old people's mobility and wheelchairs provide challenges that need to be met with information, in the product description 'we specify the number of steps up to the first floor', one told, or once an excursion bus with such guests is on its way. Standardization logics seem to have to accommodate the wish to display local peculiarities.

Cruise production also affects service quality. The rather strict and detailed quality lists from the cruise operators are expected to be met. If not they are certain to be told so. As one said, 'before the bus is back on the quay'. And any lack of information that the excursion salesperson on board can give the customer may form a basis for complaints and have repercussions for the company in question. Complaints must be dealt with promptly. The challenge, however, is to get employees with the right attitude. One restaurateur states, 'it is demanding for those who have contact with the guests ... We walk with the guests and talk about the food', because 'it is more important to talk with the guests than to clear away the plates'. But there is satisfaction in serving 'death hungry people who appreciate local food'.

Cruise volumes can be re-negotiated. When relations with cruise operators have ripened over some years, this seems to open possibilities in bargaining that allow a more powerful voice of self-assurance on the hand of the provider. One experienced restaurateur wanted to reduce stress and to reclaim more control herself. She said, 'one bus only ... I have chosen it away', even though the capacity of her establishment is bigger, seemingly opposing the law of economy of scale, but 'it is the right for me', she wants to give the smaller-group-kind of service.

Logistics and capacity are keys to excursion production. Lunches are planned according to given capacities, while other experiences have different constraints. The villages as cultural landscapes are narrow, habitable spots between sea and mountain, and quickly become crowded with buses and a multitude of spectating tourists. One interviewee says, '... it is like a New York fair', but claims that tourists think that 'Lofoten is wilderness however crowded it is'. The smaller museums of the fishing villages have few options for physical expansion. The result rarely consists of peaceful strolling in quaint surroundings. According to a guide the tourists are 'let loose with the message to be back in the buses in twenty or thirty minutes'. Some museum producers deny the regime of cruise standards and cling to the control of the guiding, wanting to present the storyline themselves. This exceeds the time slot that otherwise might be allotted. Other situations where volume makes a problem are in less standardized experiences where individuals and their skills represent the attraction core and communication with customers is essential, like smiths and glassblowers. One tells of how instead of playing up to what could be interaction with the busy cruise tourists, he concentrates on his tools and shuts out the madding crowd, waiting for the hordes to leave. But he also needs the money, so he endures.

Demands on capacity and standards and the tight schedules of excursions make hard priorities and may become impossible to meet. An ecological farm which is a tourist attraction for bus groups and individual tourists had to terminate negotiations with cruise operators. The farm did not want to expand – and the operators were not willing to give them the price they think they needed: 'we're not like Lofotr [Viking Museum]; they can lower the price and catch up through volume'. Their farm shop and toilet facilities were also below the usual standards: 'OK, it is a bit worn, but they should meet the real life … it is a farm'. Neither is there time for more than a short stop at a nice wooden church – located with a mighty view of the open Atlantic – where the organist would be more than willing to present some music. The time needed to park six buses and for slowly moving passengers to walk around, sit down and listen, experience the atmosphere and return to the bus is simply not there. 'Cruise tourists have so little time', the organist laments.

Some experiences escape the logic of bus numbers and specified arrivals/ departures, demonstrating that there is room for customization. The cruise manual lists an ocean-side golf course half an hour away from the port, a possibility open to enthusiasts to get away from the multitude of fellow passengers. On-the-spur experience making also happens. A local fisherman reported having taken cruise passengers fishing on several occasions. They walk off the cruise ship at the quay and pass on to the fishing boat at the floating dock. Information on such a possibility is available, tourists phone direct from on board or ask in the tourist kiosk on the quay, but the option does not figure in the manual. Cruise industry standard specifications have no authority in these circumstances. As the fisherman said, 'on board I am the skipper, they must behave according to my rules'.

Many respondents seem to have had a rather steep learning curve in dealing with the cruise operators. Initially there were temptations in the foreseeable number of guests and the accompanying turnover. The reminiscences have a taste of historical innocence or naivety from the early phase of cruise tourism, but several had seen that 'we were put up against each other' and that the operator 'would go to someone else' if the price was not settled. But they 'had learned' certain lessons. Through the price bargaining and the pressure for volume, the unpleasantness of last minute de-bookings and complaints-based repercussions they experienced during the first years, it seems that many producers gain a certain satisfaction from fixing situations. The 'knowledge generation' (Hollinshead 1999,19) through the initial seasons becomes more important when we move on to more co-operative reactions to cruise development.

Cruise Involvement Coordination

The rigidity of the excursion time table is a main standardization mechanism, and guides control the machinery of bus driving times, museum visit slots, village spectating, lunch buffet seating and photo shoot breaks. Thus, guides are of central importance for realizing the mass customization potential developed through the range of options and the variety over time in the excursion program. Since the

Lofoten Guide Service company was started in 1998, guide services in Norway have been increasingly organized, professionalized and integrated. The present Arctic Guide Service (AGS) covers operations in the three biggest cruise ports in North Norway, North Cape, Tromsø and Lofoten (http://www.arcticguideservice.com/destinations). There is further horizontal integration with other guide operations through the ownership of the operator ECS, thus also integrating vertically. The coordination has built a flexible apparatus capable of handling top traffic days: guides are moved between destinations along the coast when needs exceed local guide resources. The tentacles of cruise related coordination also have to draw on transport resources from a wide district. A bus company interviewee stated that a bus of sufficient quality was driven from Kiruna in Sweden, 500 km from Leknes.

Organizing a guides company represents professionalization in tourism, claimed to be a 'productive aspect of power' (Cheong and Miller 2000, 385) and institutionalizes local knowledge as constructive for producing Lofoten experiences. The certification obtained through their courses signals quality and gives AGS almost a monopolistic position vis-à-vis operators, and decisive power in dealing with the producers. The AGS, one claimed, may say, 'ok, if [a museum] is included, it shall last 20 minutes, otherwise it's out', based in 'knowledge of the market'. Cruise companies do not influence the contents of what the guides say but standardize the performance through the use of time/place schedules. Language proficiency and guide experience is believed to compensate for limited local knowledge for guides from outside the AGS; the guides interviewed related examples of misinformation and information given at the wrong spots by imported guides.

Another definite step in cooperative coordination was carried out in 2007 with the establishment of the Lofoten Cruise Network (LCN). For two years there had then been 'a cruise project' (Sandnes 2006, 1) running parallel to the DMO masterplan process, organizing the 25 companies involved in cruise services. The report (Sandnes 2006) stated the importance of quality improvements, voicing the main concerns of the operators: 'more certified guides, good and modern standard in transport, good service facilities in the businesses, language, service and understanding of the customers'/guests' needs for the good experience'. Also, the cruise operators wished lunch to be served in one hour at the maximum. Some producers had opined that the tourists needed more than 45 minutes but the report stated 'the attractions must to a greater extent … adapt the product to the time we have today' (Sandnes 2006, 2). LCN members have adapted accordingly.

As a formal organization, LCN is led by the local entrepreneur-cum-coordinator: ship agent, former cruise operator representative and AGS managing director. Again there is a bigger regional context: LCN is part of Cruise Network North Norway and Svalbard (CNNS) (CNNS 2013). This embodiment of the coordinated interests of suppliers in cruise shore excursions strengthens land based actors and the destination in marketing activities, but it also channels the order of the industry internationally. LCN is represented at cruise workshops and industry fairs like those in Hamburg, London and Miami, occasions that renew ties between long time co-operating people and a main feature of networking in this industry (Lemmetyinen 2008). Thus, the producers are linked to the global industry's perspectives on market trends and product development, information is sieved through representatives at international gatherings, and by updating and producing the cruise manual and individual product specifications accordingly the tactics of what to sell to whom are influenced. In the destination the network is also a marker of interests separate from the rest of the tourism sector. LCN people claim that 'cruise is different' and that the network is a confirmation of this, but this might also be a legitimation of keeping up differences. 'Someone had to be a bellwether', a pioneer said with reference to the early years. Today the central participants in cruise involvement are given recognition for their work.

Some say that the network is a response to 'pressure from the outside'. As one producer observed, 'we answered by organizing ... we are getting more equal. We could not keep on as we did with [ECS] steering as they wanted to', referring to cruise operators in price bargaining putting one supplier up against the other. Thus, it has been a case of the empowering effect of power exercised. Another talks of 'the knowledge accumulated in the network', a sharing of experiences, and of 'analysing it and using it'. Some claim that it led to a common awakening regarding how to face the bargaining challenges, a collective competence improvement with elements of resistance. The knowledge acquired has its concrete form in 'that binder of his', as a network member said, referring to the manager.

The LCN leader stated that 'cruise has been dressage, in expecting the same quality in the individual businesses' – like bending former self-satisfied wills to the order of the excursion standards. And the network has been dressage 'because you learn someone else is better than you', a museum manager said. It smells of Foucauldian disciplining. In the 1980s, tourism in Lofoten was still in its infancy, a part-time business: 'all the others [tourists] were just someone who came – when they came', one producer reflected. In comparison cruises represent predictability. When these amateurish actors met with the demands of the cruise operators, they faced professionality and were expected to behave accordingly. 'Cruises set a new standard that we had to relate to', one stated. The positive and productive implication of being targets in cruise relations was re-igniting tourism development of attractions and experiences at the destination in a broader sense. Opening up old or renovated *rorbu* houses to whoever came along was no longer enough.

Whatever its dressage impact, the LCN manager says it was 'a big process to get cruise involvement accepted in tourism' and that locally, 'it has been divided'. The recognition from the travel industry has been reluctant. The manager tells how they documented their economic dependency on cruise guests for themselves: 'without them we would all be in red figures'. These facts opened the acceptance of cruise related matters into the destination strategy project (Destination Lofoten 2006; Sandnes 2006). In 2006 when Destination Lofoten produced their masterplan 'Lofoten as a destination towards 2015', cruises were originally not addressed. The DMO perspective was from Svolvær, where cruises mattered less. The cruise project group produced a report themselves to demonstrate the importance of cruises. This four page note (Sandnes 2006) was included as attachment to the masterplan. The effort may have had an impact: since 2011 the DMO has taken over the secretary functions of the LCN. The strategy process illustrates the DMO power in defining the agenda and controlling access to the discussions of the plan (Marzano and Scott 2009, cf. Lukes 2005) and of the attention traditionally paid to hotel capacities and traffic access bottlenecks. Relating to cruise production demands had a mobilizing effect on LCN members, but the activity thus created also affected the wider understanding of the destination challenges.

Satisfaction, Reluctance, Avoidance and Resistance

Standardization in the Fordist sense of having preplanned numbers of tourists following a few varieties of the same programs, stopping at the same views and eating the same food at the same times is definitely an implication of cruise involvement. But the scale of cruises has not streamlined all local suppliers, wiped out originality and levelled out individual qualities, and available options compensate for inflexibility in standards. Adapting to the regime of excursion timing schedules seems to be the fate of the biggest producers – some restaurants, the largest museum, one picturesque fishing village – and where calculability logics dominate. For others the adaptations to cruise demands are met with reluctance, but they compromise on quality and welcome the income. For a few own values or having experienced the stress overshadows the profit potential. And some do not adapt at all because they are not let in, or only when they choose themselves.

Consistent service quality is a recurring theme also for cruise production (Véronneau and Roy 2009) and the quality standards presented and terms negotiated in Lofoten, in addition to cruise companies' own guest questionnaires measurements, form a control system accepted by excursion providers and through which the industry can check that passenger satisfaction is maintained. Standardization has succeeded on the planning level and in the attitudes of the core members in the LCN. A problem recognized lies with keeping up the seasonal staff quality (cf. Testa and Sipe 2012, 654) and their attitudes to guests, not with bending to cruise standards per se – because these have improved destination quality in general.

But cruise adaptation has not come without friction. A feeling of initially having been outmanoeuvred by cruise representatives' bargaining skills led to ascertaining common interests and organizing the LCN, thus recreating a feeling of being more in control of own fate and painting a more balanced and self-assured picture of the power process. Still, there are voices that call for a more confronting and harder bargaining line, partly linked to a widely shared concern for the sustainability of cruise growth. It seems the network is not the forum for solving the strategic issues concerning how to stagger the pressure of volumes. Good relations have been established between agents and targets, and cruise volumes rolling in is a sign of business success.

Lastly and contrafactually: what if the cruise industry demands had not occurred? Would production have developed in a similar professional direction? A long time view on tourism development in Lofoten must contextualize cruise impacts: Several changes converge. A new generation, educated and self-confident, is taking over some of the businesses with a willingness to work professionally, for more seasons and for longer periods of employment. They have a general awareness of the necessity to move on from 'harvesting-tourism'. Influenced by the market's wishes – through the cruise industry – consolidation was triggered, with LCN cooperation focusing on the sharing of information and experience and long-time horizons. Instead of restating the outdated view on Lofoten destination development where municipalities compete, LCN members have accepted the tourist perception of Lofoten as one destination region. Adapting to cruise tourism has been the key.

References

Bezic, H. and Vojvodic, K. (2010), 'Towards Improving Destination Port Status', *Tourism & Hospitality Management* May 2010 Supplements, 17–32.

Bærenholdt, J.O., Haldrup, M., Larsen, J., Urry, J. (2004), *Performing Tourist Places* (Aldershot: Ashgate).

Charlier, J.J. and McCalla, R.J. (2006), 'A Geographical Overview of the World Cruise Market and its Seasonal Complementarities', in Dowling (ed.).

Cheong, S.-M. and Miller, M.L. (2000), 'Power and Tourism. A Foucauldian Observation', *Annals of Tourism Research* 27:2, 371–390.

Chin, C.B.N. (2008), *Cruising in the Global Economy. Profits, Pleasure and Work at Sea* (Aldershot: Ashgate).

CNNS (2013), <http://cnns.no/cnns-members>, accessed 26 April 2013.

Coles, T. and Hall, C.M. (eds) (2008), *International Business and Tourism. Global Issues, Contemporary Interaction* (Abingdon: Routledge).

Cooper, C., Scott, N., Baggio, R. (2009), 'Network Position and Perceptions of Stakeholder Importance', *Anatolia* 20:1, 34–45.

Cruise Norway (2013), <http://www.cruise-norway.no/viewfile.aspx?id=3579>, accessed 6 May 2013.

Cruisemarketwatch (2013), <http://www.cruisemarketwatch.com/articles/cruise-market-watch-announces-2013-cruise-trends-forecast/>, accessed 9 April 2013.

Datamonitor (2011), Carnival Corporation & plc. Company Profile 18 May 2011. <www.datamonitor.com> (homepage), accessed 12 May 2012.

Destination Lofoten AS (2006), 'Lofoten som reisemål mot 2015. En Masterplan for arbeidet med bedre reisemålsutvikling i Lofoten' [Lofoten as a destination towards 2015. A masterplan for improving destination development in Lofoten] (Svolvær: Lofotrådet).

Douglas, N. and Douglas, N. (2001), 'The Cruise Experience', in Douglas et al. (eds).

Douglas, N. et al. (eds) (2001), *Special Interest Tourism: Context and Cases* (Brisbane: Wiley).

Dowling, R.K. (ed.) (2006), *Cruise Ship Tourism* (Wallingford: CABI).

Farstad, E. and Dybedal, P. (2012), 'Tenk om cruiseturistene kom med fly, buss eller bil ... '. [What if the cruise tourists came by plane, bus or car ...], *Samferdsel* 10:4–5, 2012. <http://samferdsel.toi.no/article31543–1345.html>, accessed 3 April 2013.

Ford, H. (1922), *My life and work* <http://www.gutenberg.org/dirs/etext05/hnfrd10.txt.>, accessed 21 March 2013.

Ford, R.C., Wang, Y. and Vestal, A. (2011), 'Power Asymmetries in Tourism Distribution Networks', *Annals of Tourism Research* 39:2, 755–779.

Foucault, M. (1977), *Discipline and Punish: The Birth of the Prison* (New York: Vintage Books).

Foucault, M. (1980), 'Two Lectures', in Gordon (ed.).

Gordon, G. (ed.) (1980), *Power/Knowledge: Selected Interviews and Other Writings 1972–1977* (New York: Pantheon).

Hall, C.M. (1994), *Tourism and Politics: Policy, Power and Place* (Chichester: John Wiley).

Hall, C.M. and Lew, A. (eds) (1998), *Sustainable Tourism: A Geographical Perspective* (New York: Longman).

Hobson, P. (1993), 'Analysis of the US Cruise Line Industry', *Tourism Management*, 14:6, 453–462.

Hollinshead, K. (1999), 'Surveillance of the Worlds of Tourism: Foucault and the Eye-of-Power', *Tourism Management* 20:1, 7–23.

Hultman, J. and Hall, C.M. (2011), 'Tourism Place-Making. Governance of Locality in Sweden', *Annals of Tourism Research* 39:2, 547–570.

Ionannides, D. and Debbage, K.G. (1998), 'Neo-Fordism and Flexible Specialization in the Travel Industry. Dissecting the Polyglot', in Ioannides and Debbage (eds).

Ioannides, D. and Debbage, K.G. (eds) (1998), *Economic Geography of the Tourism Industry* (London: Routledge).

Jaakson, R. (2004), 'Beyond the Tourist Bubble? Cruiseship Passengers in Port', *Annals of Tourism Research* 31:1, 44–60.

Jordan, A.E. (2008), 'Cruise Ship Pioneers. Remembering the Innovative Vessels that Changed the Face of Modern Cruise Travel', *Cruise Travel Magazine* 29:4, 26–31.

Klein, R.A. (2012), 'Different Streams in Cruise Tourism Research: An Introduction', *Tourism* 60:1, 7–13.

Kwortnik jr., R.J. (2006), 'Carnival Cruise Lines. Burnishing the Brand', *Cornell Hotel and Restaurant Administration Quarterly* 47:3, 286–300.

Lemmetyinen, A. (2009), 'The Coordination of Cooperation in Strategic Business Networks. The Cruise Baltic Case', *Scandinavian Journal of Hospitality and Tourism* 9:4, 366–386.

Lukes, S. (2005), *Power: A Radical View* (Second Edition) (Basingstoke: Palgrave Macmillan).

Lumson, L. and Page, S. (eds) (2004), *Tourism and Transport: Issues and Agenda for the New Millennium* (Amsterdam: Elsevier).

Marshall, C. and Rossman, G. (1995), *Designing Qualitative Research* (Thousand Oaks, CA: Sage).

Marzano, G. and Scott, N. (2009), 'Power in Destination Branding', *Annals of Tourism Research* 36:2, 247–267.

Miller, R.K. and Washington, K. (2010), *Travel and Tourism Market Research Handbook* (Loganville: Richard K. Miller & Associates).

Milne, S. (1998), 'Tourism and Sustainable Development. Exploring the Global-Local Nexus', in Hall and Lew (eds).

Papathanassis, A. and Beckmann, I. (2011), 'Assessing the 'Poverty of Cruise Theory' Hypothesis', *Annals of Tourism Research* 38:1, 153–174.

Patton, M.Q. (2002), *Qualitative Research and Evaluation Methods* (Thousand Oaks, CA: Sage).

Reed, M.G. (1997), 'Power Relations and Community-Based Tourism Planning', *Annals of Tourism Research* 24:3, 566–591.

Ritzer, G. (2002), *McDonalization.The Reader* (Thousand Oaks: Pine Forge Press).

Saarinen, J. (2004), 'Destinations in Change': The Transformation Process of Tourist Destinations', *Tourist Studies* 4:2, 161–179.

Sandnes, R. (2006), *Vedlegg 7 Cruise*. Attachment to Lofotrådet/Destination Lofoten 2006.

Scott, N., Cooper, C. and Baggio, R. (2008), 'Destination Networks. Four Australian Cases', *Annals of Tourism Research* 35:1, 169–188.

Sheehan, L.R. and Brent Ritchie, J.R. (2005), 'Destination Stakeholders. Exploring Identity and Salience', *Annals of Tourism Research* 32:3, 711–734.

Testa, M.R. and Sipe, L. (2012), 'Service-Leadership Competencies for Hospitality and Tourism Management', *International Journal of Hospitality Management* 31:3, 648–658.

Torres, R. (2002), 'Cancun's Tourism Development from a Fordist Spectrum of Analysis', *Tourist Studies* 2:1, 87–116.

Véronneau, S. and Roy, J. (2009), 'Global Service Supply Chains: An Empirical Study of Current Practices and Challenges of a Cruise Line Corporation', *Tourism Management* 30:1, 128–139.

Weaver, A. (2005), 'The McDonalization Thesis and Cruise Tourism', *Annals of Tourism Research* 32:2, 346–366.

Weaver, A. (2008), 'When Tourists Become Data: Consumption, Surveillance and Commerce', *Current Issues in Tourism* 11:1, 1–23.

Weaver, A. (2011), 'The Fragmentation of Markets: Neo-Tribes, Nostalgia and the Culture of Celebrity: The Rise of Themed Cruises', *Journal of Hospitality and Tourism Management* 18:1, 54–60.

Weaver, A. and Duval, D.T. (2008), 'International and Transnational Aspects of the Global Cruise Industry', in Coles and Hall (eds).

Wie, B.Y. (2005), 'A Dynamic Game Model of Strategic Capacity Investment in the Cruise Industry', *Tourism Management* 26:2, 203–217.

Wood, R.E. (2000), 'Caribbean Cruise Tourism. Globalization at Sea', *Annals of Tourism Research* 27:2, 345–370.

Wood, R.E. (2004), 'Cruise Ships: Deterritorialized Destinations', in Lumson and Page (eds).

Chapter 11

Transforming Visions and Pathways in Destination Development: Local Perceptions and Adaptation Strategies to Changing Environment in Finnish Lapland

Eva Kaján and Jarkko Saarinen

Introduction

'Future vulnerability depends not only on climate change but also on development pathway'. This general statement from the Intergovernmental Panel on Climate Change (IPCC 2007a:19) manifests some of the burning issues and discussions current in the development of tourist destinations in the Global North and Global South. Many tourist destinations are vulnerable in the future to anticipated climate change and related adaptation needs are increasingly highlighted in academic literature and international and national policy documents (see ACIA 2005; Scott and Becken 2010; Becken and Clapcott 2011; Scott et al. 2012). All this has emphasised a need to search for alternative pathways in tourist destination development, which is a crucial issue in the context of the sustainability of tourism and the industry's capacity to create positive impacts on regional development and local tourism-dependent communities (see Scott 2011; Bramwell and Lane 2012; Kaján and Saarinen 2013), for example.

While the adaptation dimension has been strongly emphasised in the recent tourism and climate change literature, the need to respond to external and internal changes and be proactive and resilient, if possible, has also been noted outside the current discussions on climate change. Actually, the issue of future changes in destination development has been studied for a relatively long time within tourism research: e.g. the evolution cycle of tourist destination (Butler 1980) represents a model aiming to demonstrate future challenges in development of a tourist destination if certain limits to growth are not considered (see Saarinen 2006). Recently, flexible and proactive evolutionary approaches have evolved to consider the development perspectives of tourist destinations under increasing or deepening external and internal change factors. The dimensions of path-dependency and path-creation in tourism development have received special interest (see Saarinen and Kask 2008; Ma and Hassink 2013). These dimensions demonstrate local-global relations, contingency and self-reinforcement, i.e., local responses to internal and external processes, structures and changes (see Baláž and Williams 2005). As a dimension path-dependence refers to a situation in which the actors and their possibilities are seen as limited by existing resources and traditions which support some development pathways better than others (Williams and Baláž 2002; Saarinen in Chapter 3 of this volume). In contrast, the dimension of path-creation emphasises the potential of local actors to influence the course of destination development and break away from path-dependency if necessary (see Nielsen et al. 1995; Stark 1996).

Path-dependency and path-creation perspectives are highly relevant in current changing environments in which traditional livelihoods or economies, that are considered 'traditional', operate. While tourism is often seen as a new form of economy in peripheral areas, in Finnish Lapland it has a relatively long history. In some respects, it could be said that tourism represents a traditional form of economy in the region – at least compared with many other currently operating economies in the region. Forest industry, for example, is a latecomer compared with tourism in many parts of the region. During the history of tourism in Lapland there have been several phases in development (see Vuoristo 2002; Lähteenmäki 2006). Currently the crucial future growth and development issues are linked to globalisation of the industry and markets, internationalisation of the customer base and changing ecological and socio-cultural environments in the high north. These are interlinked issues but, especially in relation to the latter perspective, the critical questions focus on the adaptation and adaptive capacity of the region, the industry and supporting internal and external structures and tourism-dependent local communities. Thus, a key question is how the region's tourist destinations fall into or can break away from path-dependency in changing operational environments in the future, i.e. why some places become locked into negative development paths, narrowing their future growth potential, whilst other places manage to avoid the same route (Martin and Sunlay 2006, 395).

In the context of global environmental change these kinds of questions have become highly relevant in Finnish Lapland. This chapter takes a specific view of the destination development with an emphasis on the local adaptation perspectives and strategies towards global climate change. These local views to adaptation needs, and local perceptions to change per se, are crucial for tourist destinations and tourism dependent communities' survival and they are linked to the dimensions of path-dependency and path-creation in regional development (see Williams and Baláž 2000; 2002; Saarinen and Kask 2008). The important role of the responses to global environmental change highlights the current views stating that forces constructing and transforming destinations are increasingly proving to be non-local ones (Milne and Ateljevic 2001; Saarinen 2004). However, even though global climate change or global change, for example, are often seen as external and being beyond local scale, the local knowledge perspectives and actions may have a key role to play: also externally oriented processes are mediated by and their impacts and implementation are influenced by local actors (see Massey 1991; Teo and Li 2003; Saarinen in Chapter 3 of this volume).

This chapter will review the general background of tourism development in Finnish Lapland and issues of changing climate in the region with a specific focus on the two case study communities of Kilpisjärvi and Saariselkä. Based on the case study communities, which are highly tourism dependent, the aim is to look at opportunities and challenges related to tourism development and climate change. In addition the chapter will discuss adaptation responses in relation to path-creation, i.e., what kinds of pathways local community members perceive in the context of changing environment in Finnish Lapland.

Tourism Development in Finnish Lapland

As a result of a long history of tourism in Finnish Lapland the tourism industry has become an important economic sector in the region and one that is seen to have a good growth potential in the future. The significant economic role of tourism has made it also a socially and politically important issue in regional policy-making. Indeed, tourism and tourism development arguments are used as a medium for many socio-cultural, economic and land-use goals and actions on the regional and local scales in Lapland (see Järviluoma 1993). Especially the employment impacts of tourism are seen as crucial in the region. Since the late 1990s the tourism sector has provided more employment opportunities in Lapland than any other field of the economy that utilises the region's natural resources (Saarinen 2003) and it is unlikely that this situation will change in the near future.

In Finland, Lapland's total market share of all overnight stays was approximately 12% with 2,4 million overnight stays in 2012 (Official Statistics Finland [OSF] 2013) but relatively the role of tourism is much higher in Lapland (and Åland) than any other parts of the country. According to the Regional Council of Lapland (2011), tourism generated 594 million euro in 2010 in the region and provided about 4400 jobs (equivalent to full-time direct employment). The Regional Council's strategic aim is to triple the income and double the employment by 2030. This strategic aim and its outcomes will most probably make the region increasingly and highly dependent on tourism as there are no emerging major economic sectors, other than mining, on the horizon. However, the long-term future potential of the current mining boom is still unknown, which may provide more emphasis on tourism development in land-use planning in many parts of the region and northern Finland, in general.

In order to achieve the high goals of future tourism development in Lapland, the major tourist destinations such as Levi, Rovaniemi, Saariselkä and Ylläs, are designated as growth centres. These major resorts still operate mainly based on domestic demand, creating a certain level of path-dependency in development planning. In addition, many other and more peripheral destinations and areas, like Kilpisjärvi and Utsjoki, are also leaning towards tourism growth and seeking ways to attract tourists and especially international tourists interested in wilderness environments and nature-based tourism activities in the future. Although there is a path-dependency in relying on past and current core activities and infrastructures, in general, the increase in demand will most probably require more effective and innovative ways to utilise the winter (i.e. snow period), which is the peak season, but also efforts to increase the currently weak role of summer tourism in Lapland (see Regional Council of Lapland 2011; Kaján, 2013a). These 'internal' aims basically call for path-creation in destination development and the Lapland Tourism Strategy (Regional Council of Lapland 2010) highlights the need for all-year-round tourism in the region as the first priority goal. Currently, the nature of tourism is highly seasonal in Lapland. Secondly, according to the strategy, the growth should be based on efficiency and quality where tourism innovations are being actively sought and promoted. Thirdly, the development of accessibility and transportation systems and infrastructures should ensure that the growth comes increasingly from the international markets. In this respect the international customers' needs should be emphasised in marketing (and sales), which will create extra challenges, if the environmental characteristics are changing in the future. The fourth goal refers to the quality of the environment, where the tourism-related construction should be based on customer-orientated planning and with 'Lappish identity'. According to the strategy, this quality emphasis in the goals will lead the region's tourism growth towards sustainable development.

Related to the use of official statistics, it needs to be realised that they provide a limited and potentially biased view compared with what may be happening outside the statistical system in tourism. The Finnish tourism statistics only include larger units, i.e. more than 10 rooms, cottages or caravan places per enterprise, which leaves out smaller businesses. In addition, private, semi-private (e.g. time-share systems) accommodation capacity, commercial cottages and second homes are in most cases unknown as are their total numbers. Thus, in some major tourist destinations the municipality level statistics may cover only 10% of the estimated total number of bed places in the area. In addition, it is not known if the user rates of larger (i.e. statistical capacity) and smaller (i.e. outside statistics) tourism businesses and units are equally used in a relative sense or in respect to the origin (i.e. domestic versus international) of the visitors. Related to the latter aspect, it is probable that official tourism statistics capture the scale and growth of international tourism better than for domestic tourism. Therefore the official tourism statistics should be used with caution in regional settings. This challenging relation between tourism statistics and factual tourism capacity (or tourist numbers), however, is quite a common problem in many countries (see Pearce 1995, 67; Page et al. 2001, 18–20).

As indicated, the majority of the tourists in Finnish Lapland are domestic visitors on holiday. In 2001 the rate of domestic over-night visits in Lapland was 67.4% and the rate of leisure tourism was 80.9% (Tourism Statistics 2001). Currently the tourism demand in the region is still dominated by domestic visitors, although in some destinations and seasons, such as Christmas time, international customers are a rather prominent segment. During the winter season 2005, for example, over 200 direct charter flights arrived from the UK alone to Finnish

Lapland. Even though UK-based tourists form the majority of Christmas tourists in Lapland, and especially in Rovaniemi, tourists nowadays arrive from all over the world (Tervo-Kankare et al. 2013).

The growth in international tourism largely contributes to winter tourism: in the period 2007–2012 the overnights in summer tourism (May–October) in Lapland increased by about 5%, but the market share of international tourists stayed about the same (28%). However, the overnights during the winter season (November–April) increased by 6% and the market share of international tourists grew from 46% to 49% (OSF 2013). This indicates slowly growing international tourism in Lapland in winter but a relative stagnation in summer. However, recent statistics indicate that during the past few years the growth has actually slowed, most probably due to the global economic crisis (see Kajan 2013a). All in all the market share of international overnights in the region has increased from 30% to 41.5% between 1995 and 2012 (OSF 2013). In the near future domestic tourism is 'expected' to grow by 3% and international visitors by 8% (Regional Council of Lapland 2010).

In addition to the current financial crisis, which probably presents a medium-term cycle or change in tourism development, an issue that will have a long-term impact on tourism growth and the nature of the tourism industry in Finnish Lapland relates to the estimated effects of global climate change. These internal and external changes operating in destinations will influence the demand and supply structures and most probably require major adaptation strategies, including pro-active path-creation, to take place in the region and different tourist destinations and modes.

Changing Climate and Environment in Finnish Lapland

The tourist products of Finnish Lapland are largely based on nature, its attractiveness and activities taking place in natural or semi-natural environments. International research has widely recognised that especially nature-based forms of tourism are those most influenced by global climate change in the future (Hall and Higham 2005; Saarinen and Tervo 2006; Tervo 2008). However, it is also widely agreed that the tourism industry per se will face significant changes due to the changing climate (Gössling and Hall 2006; Becken and Hay 2007; Becken and Clapcott 2011). In addition, the Intergovernmental Panel on Climate Change (IPCC 2007b) has estimated that the Arctic and northern regions will probably experience the highest rate of warming compared with other regions of the world. It is estimated that within the next 100 years the Arctic temperatures will be significantly higher, rising to even 4–7 oC during winters in land areas (ACIA 2005; IPCC 2007b).

While there is rather clear consensus on the future directions of global climate change in Lapland, the impacts and implications for the region's tourism industry and products are less clear. Indeed, due to climate change, snow accumulation may decrease or temporarily increase, certain activities may disappear; and new opportunities may arise and some seasons get longer. Scenic attractions may become more or less appealing and particular nature-based activities expand or decrease depending on the local consequences of climate change (see Johnston 2006; Kaján 2013a). Thus, there is obviously a wide range of uncertainties when dealing with long-term impacts and changes but, according to Finland's National Strategy for Adaptation to Climate Change (Marttila et al. 2005), climate changes will probably shorten the winter season and lengthen either summer or the so-called shoulder-seasons, or both. However, Finnish Lapland and northern regions in general are expected to gain a competitive advantage due the potential problems of lack of snowfall in the south (see also Saarinen and Tervo 2006; Tervo 2008; Tervo-Kankare 2011).

Thus, depending on their direction and character, the expected changes in the environment can create opportunities or economic losses for the tourism industry and tourism-dependent communities in Lapland in the future. This calls for adaptive strategies and path-creation approaches on local and regional scales. However, while the presence of climate change has been recognised in regional tourism policies, such as in the previous Lapland Tourism Strategy 2007–2010 (Regional Council of Lapland 2008), the adaptive elements – in contrast to mitigation issues – are not coordinated in practice by national or regional authorities and policy-makers. The above-mentioned Finland's National Strategy for Adaptation to Climate Change (Marttila et al. 2005) gives an overview of the issues related to adaptation needs and general strategies, but the implementation seems to be delegated to regions and operators. Therefore, it is relevant and justified to analyse how the issues of environmental and climate change are considered in a local scale and what kind of challenges and responses in adaptation exist among local actors.

The following study is based on 47 interviews which were conducted in 2011–2012 in two communities of Kilpisjärvi and Saariselkä in Finnish Lapland. The thematic open-ended questions concerned the current climate-related changes and tourism development in general. The interviewees represented a wide range of community members who live permanently in the area. The interviewees were owners or employees of different establishments or institutions and had other livelihoods. In addition, some interviewees were retired. Though the main results are also discussed in Kaján (2013;2014), this chapter gives a more detailed overview of the changes observed by the communities and their adaptation responses and challenges in respect to transforming visions and pathways in tourist destination development.

Case Studies of Saariselkä and Kilpisjärvi

Brief historical account

The selected two case study communities and destinations were chosen based on their location in the Arctic, the existence of tourism and the preconceived differences

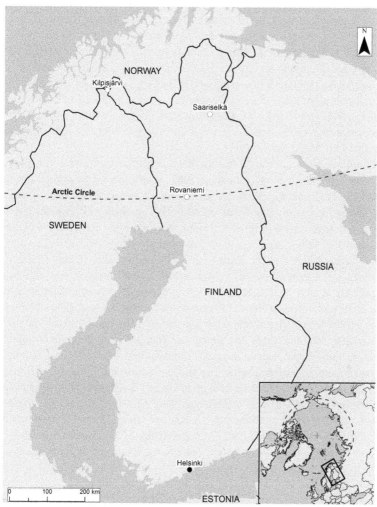

Figure 11.1   Locations of Kilpisjärvi and Saariselkä case study sites
Source: Authors

in their levels of tourism dependency. Saariselkä is a touristic community created by the evolution of tourism, while Kilpisjärvi has also other existing livelihoods and history beyond touristic use of the environment. Kilpisjärvi and Saariselkä are both situated north of the Arctic Circle in Finnish Lapland (Figure 11.1).

Both areas have for centuries been used by the indigenous Sami population in reindeer herding and other traditional livelihoods. The permanent settlement in Kilpisjärvi dates back to the 1910s and the first inn was also built there at that time. The Finnish Tourist Agency established a hostel in 1937 and the first hikers began to arrive at what is currently known as Finland's most popular wilderness environment, Käsivarsi. The area is dominated by Saana and Malla Fells. Saana became a protected area in 1916 and Malla was designated as a strict nature reserve in 1938. Several other protected areas were established in Lapland in that year, which again increased tourism in the region. The era marks the beginning of a nation-wide interest in hiking and cross-country skiing as recreational activities. The road and telephone lines arrived to Kilpisjärvi in 1942 as a result of World War II and the National Border Guarding Forces established a station near the village in 1945. The 1940s were a busy time in Kilpisjärvi in terms of infrastructure development and the construction work for a hotel, which is still in operation, began in 1949 (Metsähallitus 2011).

The unique flora and fauna has attracted scientists to the area for long time: the Biological Station was opened in 1963, providing jobs for the community (University of Helsinki 2004). A milestone in Kilpisjärvi's history occurred in 1978 when a grocery store was opened. This marked an important beginning in trans-boundary trade with Norwegians, which still dominates village life. Kilpisjärvi was fully connected to the national electricity grid in 1982 and as the school was established in the same year it allowed families to stay in the village permanently (Metsähallitus 2011). According to the interviewees, this marked a significant change in locals' lives, combined with permanent housing plans, which enabled families to rent land and build private housing. The planning and building rules were strict at that time and the housing had to match the scenery. Land-use planning has since developed and enabled more permanent housing as well as building businesses, but such strict policies concerning landscaping and construction are not enforced anymore and are a constant source of discontent.

A rapid expansion of tourism took place in the 1990s when the Holiday Centre (Kilpisjärven Lomakylä) was established. This allowed Norwegians to place 40 caravans in Kilpisjärvi permanently – a concept which has evolved to about 300 caravan places in the village (2013), and contributes significantly to village life. The combination of the surrounding protected areas and unique landscape, a variety of administrative jobs, trans-boundary trade, and the increasing tourism, have all contributed to the characteristics of Kilpisjärvi today.

The initial history of tourism in Saariselkä is linked with gold panning. After the major gold boom came to an end, many of the prospectors' buildings accommodated hikers and skiers. The Laanila Inn, for example, was established in 1912 and it was the first tourism service in what is now known as Saariselkä. Before that, from 1902 to 1912, the house served as the main office building for the Aktiebolaget Prospector mining company (Saarinen 2001).

An increase in the numbers of hikers in the mid-1950s caused problems that are still well known today: littering, erosion, path and hut maintenance and conflicts between livelihoods (Partanen 1992). After World War II, the reconstruction of tourism facilities in Lapland lasted until the 1960s which also signified a turn in consumer demand. The nearby Urho Kekkonen National Park was established in 1983 as a result of a rapidly expansion of hiking in the late 1960s (Partanen 1992) but the demand was also increasingly focused on activities such as downhill skiing. As a result, the first major ski resort in Lapland was established in Saariselkä with a wide range of services. A spa, indoor activity centre, pizzeria, grocery stores and boutiques, nightclubs, restaurants, a brewery etc. were developed (Saarinen 2001). In addition to the main resort, other minor tourist sites in the destination were developed, such as the resorts of Kiilopää and Kakslauttanen. The history of gold panning in the region was used in relation to the development of Tankavaara resort, near the village of Vuotso south of the main resort. In Tankavaara, there is a gold museum and a reconstructed village representing the time of the Finnish (and somewhat surprisingly North American) gold rush in the late nineteenth and early twentieth centuries.

Recent reorientations (development strategies and pathways)

Currently, Kilpisjärvi has 148 inhabitants, and 21 businesses that are for the most related to tourism such as accommodation providers, retail or programme services. Saariselkä has a permanent population of 341, with 66 businesses with similar tourism-related sectors (Statistics Finland 2012). There seems to be a strong tourism-related entrepreneurship in both communities. Kilpisjärvi had 1400 bed places in 2009, and according to a development plan the aim is a growth of almost 50% by 2020 (Municipality of Enontekiö 2010). The portion of foreign visitors in Kilpsijärvi is 18% which is much below Lapland's average of 35%. In Saariselkä the situation is reversed, with an annual average of foreign overnights of 46%, although during peak seasons (e.g. Christmas) the proportion is as much as 75% (Municipality of Inari 2008; Municipality of Enontekiö 2010; Regional Council of Lapland 2011).

The local development strategy of Saariselkä states that the aim is to increase the 13 500 registered and unregistered bed places in 2007 to 40 000 in 2020 (Municipality of Inari 2008). The overnights in winter (November–April) have increased by 22% from 2001 to 2012 whereas overnights in summer have increased by just 11%, signifying the growing importance of winter tourism (Regional Council of Lapland 2011). Urho Kekkonen National Park, which adjoins Saariselkä, received 277 000 visitors in 2011 and similarly the protected areas surrounding Kilpisjärvi received 74 500 visitors (Metsähallitus 2012) indicating the importance of nature-based tourism in the area. Nowadays both destinations have extensive hiking and cross-country skiing networks and tourism in both locations is highly nature-based. They are also considered as leading tourist destinations in their respective municipalities (Kaján 2013a). However, Saariselkä's development vision goes far beyond the local and regional scales, stating that the place aims to be the leading nature-based tourism destination in Europe by 2015. While both destinations are still largely based on path-dependency in their development orientation, there are discussions on alternative development routes. Next, local perceptions and responses to estimated environmental and climate change impacts are reviewed with a view to possible elements of path-creation by local actors.

Local perceptions of change

Based on the interviews, the *development-related* opportunities and challenges in Kilpisjärvi and Saariselkä were related to nature-based tourism and the increasing role of protected areas in the future development (Table 11.1). In both communities the interviewees considered strong seasonality as one of the existing development challenges. As both communities are strongly dependent on snow-based activities, changing climate may cause negative effects on the important peak season. The current traditional clientele in both communities was also considered as being the ageing repeat cross-country skiers who may have negative views to some of the relatively new customer segments and evolving tourist activities (Figure 11.2). This ageing customer-base was perceived as one of the current development problems. In addition, practices and actions towards developing infrastructure related to trails and skiing tracks in wilderness and protected areas were also considered challenging, though this was linked to the opportunities that arose from the increasing role of nature-based tourism.

Table 11.1   Development-related opportunities and challenges as perceived by the community members (based on Kaján 2014)

	Kilpisjärvi	Saariselkä
Opportunities	Role of nature in tourism development Increasing role of Norwegians More emphasis on authenticity Increasing role of internet Cultural tourism 'Turning point' in tourism development	Role of nature-based tourism Increasing role of international tourists Air access Diversification Good infrastructure Variety of services Staying small and unique
Challenges	Conflicts between livelihoods More investment in path- & skiing-networks needed Land –use planning issues Ageing customer-base Long distances Difficult to access Seasonality Lack of services Weak infrastructure (water treatment and electricity)	Developing national park Ageing customer-base Poor marketing Pressure to expand Local politics Power imbalance Dependent on air arrivals High property prices Lack of product development Seasonality

Figure 11.2   Snowmobiling outside the official routes on Lake Kilpisjärvi. Since the mid-2000s recreational snowmobiling has grown in the region creating increasing infrastructure and service needs and critical land-use discussions among local community and more 'traditional' tourist groups preferring non-motorised activities

Among the interviewees the increasing international tourism was widely considered to open new possibilities. However, the internationalisation aims were seen as challenged by accessibility (transportation) options and development possibilities. The interviewees also indicated that balancing between being small and attractive with development and land-use pressures generates challenges.

Considering the current role of nature in local livelihoods through tourism, climate change may have severe effects on local communities. Questions about the opinions concerning climate change 66% (N=31) indicated that some kind of change was indeed occurring. These perceived changes were either seen as anthropogenic or caused by natural cycles, whereas a minority (14.5%, N=7) of respondents did not believe in climate change at all (Table 11.2). The remaining respondents, 19.5% (N=9), were unsure about the phenomenon and its existence.

Climate change can be considered to occur through extreme weather events, temperature variations and warming (IPCC 2007b) and all three elements were perceived to be already present in the case study communities. This is a rather important issue in respect to adaptation strategies and pathway creation: if local communities fail to identify threats or opportunities related to climate change, they will have no reason to respond to those risks or new prospects, or to build

Table 11.2   Community attitudes towards climate change (N=47)

	Yes	No	Can't say	Total (N)
Saariselkä	17	5	4	26
Kilpisjärvi	14	2	5	21
Total	31	7	9	47

adaptive capacity (see Davidson et al. 2003). When asked about *climate change-related* opportunities and challenges, the interviewees indicated a fairly coherent perception about its positive impacts (Table 11.3). Relatively, snow security was the most dominant factor contributing to positive impacts followed by warming summers. Negative effects generated far more diverse and dispersed answers such as effects on winter tourism with temperature variations and shorter permanent snow cover (see Kaján 2013b).

Table 11.3   Climate change –related perceived impacts (modified from Kaján, 2014)

Positive impacts	Negative impacts
Snow security Warmer summers Increasing snowfall Security compared with other regions Forces new ways of thinking Increased interest in unique places Increased awareness of climate change	Shorter winter season Permanent snow later Longer shoulder seasons Decrease in snow days and snow cover Rain in winter Weakening ice sheet Decreasing permafrost Great temperature variations in the spring, Increased blizzards **Extremes:** Increased extreme temperatures Single extremes (impacts on image) **Others:** Increased winds Loss of unique habitat Problems in tourist generating countries Summers too warm Increasing temperatures Changing landscape: increase in pine trees More insects: ticks and horseflies Earlier arrival of mosquitos Challenges in infrastructure

Local responses in adaption and issues of path-creation

The challenges related to development and climate change have generated a variety of adaptive actions, though the reasons behind these vary. These actions seeking new development pathways, i.e. path-creation elements, could be divided into diversification and technology-related adaptation responses to increase coping range within seasons and decrease seasonality itself (see Kaján 2013). In general, they can also be seen as actions aiming to avoid lock-in situations in development and changing environment.

Based on the interviews, diversification processes exist to combat strong seasonality and lessen the snow-dependency. The diversification includes product development *within* seasons as well as emphasising *new* seasons. Increasing summer tourism was seen as one of the most dominant alternatives to decrease snow-dependency. In addition, introducing indoor winter activity settings, non-snow dependent activities (i.e. viewing northern lights) and events were also mentioned. Product development may be a potentially effective way for adaptation and path-creation. The diversification process requires considerations of creating a new customer-base, as new products may not necessarily attract the old clientele. Despite the risks of losing current segments, creating new customer-bases was actively sought for in both communities (see Kaján 2013, 2014).

Technology-related adaptation was more evident in Saariselkä than in Kilpisjärvi. The key element of technology-related adaptation was the development of artificial cross-country skiing tracks. This adaptation strategy is established to provide a more reliable winter season-start and is currently utilised in many other places. It is also a widely used adaptation tool in down-hill skiing environments, where it represents a path-dependency oriented technique in development, which is based on the resort model in Saariselkä from the 1970s and 1980s. At that time the development thinking aimed to focus on mass scale tourism products and activities (see Saarinen 2001). Recently the previous development models have been revised towards nature-based tourism activities (Saarinen 2007) with the mentioned new vision of Saariselkä: 'Europe's leading nature tourism destination'.

Though snow-making is a well-known adaptation method (see Scott et al. 2006; Moen and Fredman 2007; Scott and McBoyle 2007; Pickering and Buckely 2010; Morrison and Pickering 2013), its role in cross-country skiing context hasn't been largely explored. In this respect, the artificial cross-country skiing tracks can support the new development avenue and function as a tool for path-creation at destination level. In addition, web cameras and the internet in general were found to be beneficial in communicating the current conditions to the consumers (Kaján 2013) but there were no major processes or elements referring to an actual search for new avenues of development based on them. Rather, their utilisation was focused more on supporting the existing development paths and modes.

Discussion and Conclusions

Tourism has become an important policy tool dedicated to change and development in Finnish Lapland. Though winter is currently the main tourism season, summer has a gradually increasing role and there is a great deal of emphasis on all-year-round tourism development. The future growth with limited seasonality is largely to be based on expanding foreign markets and on nature-based activities. The future of nature-based tourism in relation to climate change seems to be determined by interdependent factors such as policy-making, seasonality, forms of tourism and intensity of external and internal environmental changes (see Hall and Higham 2005), issues which in the end are highly contextual and place-specific. Indeed, the perceived climate change impacts on the tourism industry differ between regions and even actors (Agnew and Viner 2001; Patterson et al. 2006; Saarinen and Tervo 2006),

   This chapter has not focused on the historical transformation or transition process of destinations *per se* (see Williams and Baláž 2002; Coles 2003; Saarinen and Kask 2008) but explored the ways in which local communities perceive, experience and aim to respond to changes related to tourism development and climate in relation to finding new pathways. In addition, the chapter also examined local adaptive responses to the predicted and occurring changes. While the emphasis has been on current perceptions, experiences and responses it is important to note that they do not happen in a socio-cultural or historical vacuum. On the contrary, the histories of Kilpisjärvi and Saariselkä play an important role and also indicate that the built environment and infrastructures in both communities have been closely linked with tourism development (see Saarinen 2001), which can be considered to have influenced the currently evident path-dependency with tourism growth. This is a similar kind of situation to that presented in the study by Ma and Hassink (2013, 101) in the Gold Coast, Australia, where initial tourism resources, conditions and structures have 'shaped the environment' for a regional tourism development path in which both path-dependency and path-creation are taking place.

   In Kilpisjärvi and Saariselkä, recent orientations indicate that both destinations are increasingly relying on nature-based tourism, which represents a traditional and obviously a realistic resource for them and many other northern and Arctic destinations (see Viken 1995; Tervo-Kankare 2011), for example, but if relied solely on nature it can make them highly vulnerable to estimated climatic changes in future. The challenges related to climatic changes included unpredictable weather in winter and challenges in infrastructure and accessibility. Temperature variations and single extreme weather elements were also considered to generate potential challenges. In return, the benefits included snow-security and better conditions during summer season. Summer tourism, which has been a challenge for fairly long time in tourism development in Lapland, may indeed benefit from climate change.

   Adaptation responses have the power to alter the pathways in development, i.e., help to break away from a negative path-dependency in changing internal and external operational environment. In ideal cases, adaption can decrease the development and climate-related negative impacts and, thus, grasp new opportunities and create a platform for co-evolutionary development which would involve tourism businesses, community groups, regional and governmental institutions and other public organisations (see Ma and Hassink 2013). However, current adaptive actions in Saariselkä and Kilpisjärvi are mainly operating in a micro scale and linked with technology and seasonal and product-related diversification processes in the tourism industry. These efforts are not only related to climatic changes but also challenges generated by current modes of tourism development representing path-dependency but the efforts are not intersectional or multi-scalar and the destination development modes were still mainly based on path-dependent dimensions. A better utilisation of protected areas was seen as an opportunity along with internationalisation. However, increasing the role of nature in the expense of other potential resources for tourism can also cause these destinations to be more vulnerable in the future, especially if the worst case scenarios of global climate change occur on a local scale.

Though tourism continues to have a strong hold in local and regional economies, its characteristics may be further developed through diversification with the aid of new technologies or new ways of using current technology. Thus, the climate-induced changes may potentially act as catalysts for new path-creation within local and regional tourism development. This requires an active search for alternative development paths that would involve local communities and cultures, for example, to tourism operations and products. This would also turn local tourism-dependent communities into subjects in regional development and empower them to be more actively involved with the most probably needed path-creation in tourism in future. Rather than seeing climate change solely as a direct threat to northern communities, it may – in some circumstances and time scales– prove to be a tool to release them from old habits and force new innovations in development. Obviously, while the tourism industry may potentially directly benefit from changing climate in the future if adaptive responses are successful, the wider internal and external structures of tourism-dependent local communities will greatly influence the course of their development pathways. Thus, answering the key question focusing on how tourist destinations can avoid lock-in situations and break away from path-dependency needs to take into account also the wider changing operational environment and relations in the local-global nexus.

References

ACIA (Arctic Climate Impacts Assessment) (2005), *Impacts of a Warming Arctic: Arctic Climate Impacts Assessment* (Cambridge: Cambridge University Press).
Agnew, M.D. and Viner, D. (2001), 'Potential Impacts of Climate Change on International Tourism', *International Journal of Tourism and Hospitality Research* 3:1, 37–60.

Baláž, V. and Williams, A. (2005), 'International Tourism as Bricolage: An Analysis of Central Europe on the Brink of European Union Membership', *International Journal of Tourism Research* 7:2, 79–93.

Baum, T. (1998), 'Taking the Exit Route: Extending the Tourism Area Life Cycle Model', *Current Issues in Tourism* 1:2, 167–175.

Becken, S. and Clapcott, R. (2011), 'National Tourism Policy for Climate Change', *Journal of Policy Research in Tourism, Leisure and Events* 3:1, 1–17.

Becken, S. and Hay, J.E. (2007), *Tourism and Climate Change: Risks and Opportunities* (Clevedon: Channel View Publications).

Britton, S.G. (1991), 'Tourism, Capital, and Place: Towards a Critical Geography of Tourism', *Environment and Planning D: Society and Space* 9:4, 451–478.

Butler, R. (1980), 'The Concept of a Tourist Area Cycle of Evolution: Implications for Management of Resources', *Canadian Geographer* 24:1, 5–12.

Coles, T. (2003), 'Urban Tourism, Place Promotion and Economic Restructuring: The Case of Post-Socialist Leipzig', *Tourism Geographies* 5:2, 190–219.

Davidson, D.J., Williamson, T. and Parkins, J.R (2003), 'Understanding Climate Change Risk and Vulnerability in Northern Forest-based Communities', *Canadian Journal of Forest Research* 33:11, 2252–2261.

Gössling, S. and Hall, C.M. (2006), 'An Introduction to Tourism and Global Environmental Change', in Gössling and Hall (eds).

Gössling, S. and Hall, C.M. (eds) (2006), *Tourism and Global Environmental Change: Ecological, Social, Economic and Political Interrelationships* (London: Routledge).

Hall, C.M. and Higham, J.E.S. (2005), 'Introduction: Tourism, Recreation and Climate Change', in Hall and Higham (eds).

Hall, C.M. and Higham, J.E.S. (eds) (2005), *Tourism, Recreation and Climate Change* (Clevedon: Channel View Publications).

Hall, C.M. and Johnston, M.E. (eds) (1995), *Polar Tourism: Tourism in the Arctic and Antarctic Regions* (Chichester: John Wiley & Sons).

IPCC (2007a) Summary for Policymakers. In: Climate Change 2007: Impacts, Adaptation and Vulnerability. Contribution of Working Group II to the Fourth Assessment Report of the Intergovernmental Panel on Climate Change, M.L. Parry, O.F. Canziani, J.P. Palutikof, P.J. van der Linden and C.E. Hanson, Eds., Cambridge University Press, Cambridge, UK, 7–22.

IPCC (2007b), 'Climate Change 2007: The Physical Science Basis', in Solomon et al. (eds).

Järviluoma, J. (1993), Paikallisväestön asennoituminen matkailuun ja sen seurausvaikutuksiin – esimerkkinä Kolarin kunta, Reports 110 (Oulu: University of Oulu, Research Institute of Northern Finland).

Kaján, E. (2013), 'Community Perceptions to Climate Change in Finnish Lapland: Examining Vulnerabilities and Adaptive Responses to the Changing Characteristics of Arctic tourism', *Nordia Geographical Publication* 42.

—— (2014)', Arctic Tourism and Sustainable Adaptation: Community Perspectives to Vulnerability and Climate Change', *Scandinavian Journal of*

Hospitality and Tourism (Special Issue on Nordic Perspectives on Tourism and Climate Change).

Lähteenmäki, M. (2006), *Terra Ultima, a Short History of Finnish Lapland* (Finland: Otava).

Ma, M. and Hassink, R. (2013), 'An Evolutionary Perspective on Tourism Area Development', *Annals of Tourism Research* 41, 89–109.

Martin, R.L. and Sunlay, P.J. (2006), 'Path Dependency and Regional Economic Evolution', *Journal of Economic Geography* 6, 395–438.

Marttila et al. (2005), Finland's National Strategy for Adaptation to Climate Change. Publication 1a/2005 (Finland: Ministry of Agriculture and Forestry).

Massey, D. (1991), 'The Political Place of Locality Studies', *Environment and Planning A* 23: 267–281.

Metsähallitus (2011), *At the Edge of Scandinavia- Exhibition* (Kilpisjärvi: Customer Service Point).

Metsähallitus (2012), '*Kansallispuistojen, valtion retkeilyalueiden ja eräiden muiden suojelu- ja virkistysalueiden kävijöiden rahankäytön paikallistaloudelliset vaikutukset vuonna 2010*' (Finland: Metsähallitus).

Milne, S. and Ateljevic, I. (2001), 'Tourism, Economic Development and the Global-Local Nexus: Theory Embracing Complexity', *Tourism Geographies* 3:4, 369–393.

Moen, J. and Fredman, P. (2007), 'Effects of Climate Change on Alpine Skiing in Sweden', *Journal of Sustainable Tourism* 15:4, 418–437.

Morrison, C. and Pickering, C.M. (2013)', Perceptions of Climate Change Impacts, Adaptation and Limits to Adaption in the Australian Alps: The Ski-Tourism Industry and Key Stakeholders', *Journal of Sustainable Tourism* 21:2, 173–191.

Mose, I. (ed.) (2007), *Protected Areas and Regional Development in Europe: Towards a New Model for the 21st Century* (Aldershot: Ashgate).

Municipality of Enontekiö (2010), *Kilpisjärvi 2020, Kehittämissuunnitelma.* [Kilpisjärvi 2020 Development Plan], May 2011 (Finland: Author).

Municipality of Inari (2008), 'Saariselkä 2020. Arktisten luontopalvelujen, liikunnan ja hyvinvoinnin lomakeskus', *Saariselän yleiskaavan ja toiminnallisen kehittämisen suunnitelma.* (Finland: Author).

Official Statistics of Finland (OSF) (2013), *Accommodation statistics* [e-publication]. ISSN=1799–6325. Helsinki: Statistics Finland [referred: 14.4.2013]. Access method: <http://www.stat.fi/til/matk/tau_en.html>.

Patterson, T., Bastianoni, S., and Simpson, M. (2006), 'Tourism and Climate Change: Two-Way Street, or Vicious/Virtuous Circle?', *Journal of Sustainable Tourism* 14:4, 339–348.

Page, S.J., Brunt, P., Busby, G. and Connell, J. (2001), *Tourism: A Modern Synthesis* (London: Thompson Learning).

Partanen, S.J. (1992), *Saariselkä – magneettimäestä matkailukaupungiksi* (Finland: Suomen Matkailuliitto).

Pearce, D.G. (1995), *Tourism Today: A Geographical Approach* (Harlow: Longman).

Pickering, C.M. and Buckley, R.C. (2010), 'Climate Response by the Ski Industry: The Shortcomings of Snowmaking for Australian Resorts', *Ambio* 39:5/6, 430–438.

Regional Council of Lapland (2010), *Lapland's Tourism Strategy 2011–2014* (Finland: Author).

Regional Council of Lapland (2011), *Tourism Facts in Lapland – Statistical Review, 2011* (Finland: Author).

Saarinen, J. (2001), 'The Transformation of a Tourist Destination – Theory and Case Studies on the Production of Local Geographies in Tourism in Finnish Lapland', *Nordic Geographical Publications* 30:1, 1–105.

—— (2003), 'The Regional Economics of Tourism in Northern Finland: The Socio-Economic Implications of Recent Tourism Development and Future Possibilities for Regional Development', *Scandinavian Journal of Hospitality and Tourism* 3:2, 91–113.

—— (2004), 'Destinations in Change': The Transformation Process of Tourist Destinations', *Tourist Studies* 4:2, 161–179.

—— (2006), 'Traditions of Sustainability in Tourism Studies', *Annals of Tourism Research* 33:4, 1121–1140.

—— (2007), 'Protected Areas and Regional Development Issues in Northern Peripheries: Nature Protection, Traditional Economies and Tourism in Urho Kekkonen National Park, Finland', in Mose (ed).

Saarinen, J. and Kask, T. (2008), 'Transforming Tourism Spaces in Changing Socio-Political Contexts: The Case of Pärnu, Estonia, as a Tourist Destination', *Tourism Geographies* 10:4, 452–473.

Saarinen, J. and Tervo, K. (2006), 'Perceptions and Adaptation Strategies of the Tourism Industry to Climate Change: The Case of Finnish Nature-Based Tourism Entrepreneurs', *International Journal of Innovation and Sustainable Development* 1:3, 214–228.

Scott, D. and Becken, S. (2010), 'Adapting to Climate Change and Climate Policy: Progress, Problems and Potentials', *Journal of Sustainable Tourism* 18:3, 283–295.

Scott, D., Hall, C.M. and Gössling, S. (2012), *Tourism and Climate Change: Impacts, Adaptation & Mitigation* (London: Routledge).

Scott, D., McBoyle, G., Minogue, A. and Mills, B. (2006), 'Climate Change and the Sustainability of Ski Based tourism in Eastern North America: A Reassessment', *Journal of Sustainable Tourism*, 14:4, 376–398.

Scott, D. and McBoyle, G. (2007), 'Climate Change Adaptation in the Ski Industry', *Mitigation and Adaptation Strategies for Global Change* 12, 1411–1431.

Shaw, G. and Williams, A. (eds) (1997), *Rise and Fall of British Coastal Resorts: Cultural and Economic Perspective* (London: Mansell).

Solomon, S. et al. (eds) (2007), *Contribution of Working Group I to the Fourth Assessment Report of the Intergovernmental Panel on Climate Change* (Cambridge: Cambridge University Press).

Statistics Finland (2012), *Suomi postinumeroalueittain 2011*. [Finland by postcodes] (Finland: Statistical Databases).

Teo, P. and Li, H.L. (2003), 'Global and Local Interactions in Tourism', *Annals of Tourism Research* 30:2, 287–306.

Tervo, K. (2008), 'The Operational and Regional Vulnerability of Winter Tourism to Climate Variability and Change: The Case of the Finnish Nature-Based Tourism Entrepreneurs', *Scandinavian Journal of Hospitality and Tourism* 8:4, 317–332.

Tervo-Kankare, K. (2011), 'The Consideration of Climate Change at the Tourism Destination Level in Finland: Coordinated Collaboration or Talk about Weather?', *Tourism Planning & Development* 8:4, 399–414.

Tervo-Kankare, K., Hall, C.M. and Saarinen, J. (2013), 'Christmas Tourists' Perceptions to Changing Climate in Rovaniemi, Finnish Lapland', *Tourism Geographies* (forthcoming).

Tourism Statistics (2001), *SVT/Transportation and tourism 2001: 11* (Helsinki: Statistics Finland).

University of Helsinki (2004), *Kilpisjärven Biologinen Asema* <www.helsinki.fi/university>.

Viken, A. (1995), 'Tourism Experiences in the Arctic – The Svalbard Case', in Hall and Johnston (eds).

Vuoristo, K.-V.(2002), 'Regional and Structural Patterns of Tourism in Finland' *Fennia* 180:1/2, 251–259.

Williams, A.M. and Baláž, V. (2000), *Tourism in Transition* (London and New York: I.B. Tauris).

Williams, A.M. and Baláž, V. (2002), 'The Czech and Slovak Republics: Conceptual Issues in the Economic Analysis of Tourism in Transition', *Tourism Management* 23:1, 37–45.

Williams, A.M. and Shaw, G. (1997), 'Riding the Big Dipper: The Rise and Decline of the British Seaside Resort in the Twentieth Century', in Shaw and Williams (eds).

Chapter 12

A Hotel Waiting for Renovation: Pallas as a Challenging Case for Tourism Development in Finnish Lapland

Seija Tuulentie and Jenni Lankila

Introduction

The old wooden hotel Pallas, which is situated in the Pallas-Ylläs national park in Finnish Lapland, has been waiting for renovation for almost 40 years. Its renovation has become an extremely conflictual issue and the hotel itself a materialisation of the problematic nature of tourism destination development. Over recent decades, several other ski resorts in Finnish Lapland have been expanding and have turned into urban-like areas in the middle of the wilderness. However, certain features make the case of hotel Pallas especially complicated. Firstly, its location in a national park means in the Finnish context that it should not be there: in Finland, a national park is supposed to be an uninhabited wilderness area. Secondly, the place has a long cultural history in tourism and it has been nominated as a national landscape, which means that not only locals but also people from Southern Finland would like to have their say concerning its development.

The fate of the hotel raises many questions about tourism planning and destination development. In this chapter, we use the case of Pallas to illustrate how temporal, material and societal issues are intertwined in destination development and how tourism development is never a truly local issue. In this chapter, we analyse the case of Pallas in order to discuss overall tourism destination development in Lapland and to study different perspectives on a nationally important place with high amenity values.

The investigation in this study is based on the analysis of planning documents, media coverage, expert interviews and the village meetings arranged by the researchers in order to understand the characteristics and development of the resort from a wider perspective than just the mountain resort itself. The discussions gathered 10–20 participants each in the villages of Särkijärvi, Raattama, Kerässieppi, Jerisjärvi and Olos, which are located closest to Pallas (see Figure 12.2), at the end of 2011 and at the beginning of 2012. Most participants were locals although three of the groups included also some second home owners, and the group in the Olos resort comprised exclusively second homeowners.

On these bases, we analyse the different possible development paths both for the hotel, which has been waiting for renovation for 40 years, and for its surrounding villages with active, small-scale tourism actors. From the point of view of a small-scale resort such as Pallas with its one hotel and visitor centre, we ask what is the consequence of the strategic choice that tourism should concentrate on big resorts with strong growth orientation. In relation to the concept of path dependence, we ask whether this kind of strategic growth-oriented operation mode gives possibilities to fundamental changes in development and thus to more path-creative actions. Firstly, we describe the wider tourism development in the area where Pallas resides. Secondly, we discuss the tourist resort's developmental paths from a theoretical point of view. After that, we concentrate on the various views on the advancement of Pallas and their consequences for future development in the region.

Tourism Context and Policies in the Region

Northwest Lapland started to intrigue writers, explorers and scientists in the 16th and 17th centuries. Those who travelled to the region were mainly interested in nature, but also in the habits and ways of life of the indigenous Sámi people and other locals. This area around Tornio River was the destination for many travellers because of the relatively easy access along the rivers. It is said to have been the best-known part of Finland for people in Western Europe in the 18th and early 19th centuries (Lapinkävijät, no date). After this specific kind of early travelling, the contemporary mode of tourism started to grow in the region in the 1930s as winter sports made their breakthrough. In particular, Pallas fells came to be known as the most suitable place for cross-country and alpine skiing in Finland. The place was 'discovered' for tourism in the 1930s by a group of female skiers who were looking for a suitable place for newly invented winter sports holidays (Lapinkävijät, no date; Kari 1978; Sippola and Rauhala 1992). Consequently, the region has a long touristic history, which has also formed the basis for the distinctive image of the area.

Before the Second World War, in 1938, a hotel was built in the fell. That period was also the beginning of another new kind of use of the place: later in 1938, the area became a national park – almost 30 years after the first initiative. Thus, the new hotel came to be situated inside the conservation area – a fact that has had a big influence on the development of the place as a tourist resort. The hotel was burnt down by the Germans during World War II but it was rebuilt and reopened in 1948 (Sippola and Rauhala 1992). Since the 1960s, there have been proposals for the enlargement of the resort as well as the hotel itself. However, the hotel is still the same as it was in 1948.

Nowadays, the only buildings in the Pallas fells resort are the hotel, which was sold by a state-owned enterprise to a private entrepreneur in 1997, a national park visitor centre and some accommodation buildings for staff. In addition, there are two ski lifts and a caravan site (Nyman 2004; Ministry of the Environment 2008). The place is also an important gateway for hikers and cross-country skiers

Figure 12.1 The old wooden Hotel Pallas waits for renovation. Together with the national park visitor center it forms a small resort

Source: Seija Tuulentie

Figure 12.2 The national park visitor centre in Pallas

Source: Seija Tuulentie

heading to the Pallas-Ylläs national park, which is the most visited park in Finland with 436,000 visits in 2010 (Metsähallitus 2011).

This chapter focuses on Pallas fells as an example of nature-based tourism in the municipality of Muonio. The mountain has a long tradition as a tourist destination but according to Richard Butler's theory of a tourist area's cycle of evolution (Butler 1980), it can be regarded as being in the decline stage. In addition, compared with the ideals of national and regional tourism strategies, the development of Pallas is distinctive, and from some points of view problematic.

One of the main emphases in Finland's Tourism Strategy to 2020 is the development of big tourist resorts. The strategy points out that this should not prevent support for innovative action nor exclude small enterprises (Ministry of Trade and Industry 2006; Hakkarainen & Tuulentie 2008). The Lapland Tourism Strategy 2011–2014 (Regional Council of Lapland 2011) also states that it concentrates on development efforts that emphasise tourism resorts and tourism zones. The zone idea comes from the rural development plan (Regional Council of Lapland 2009), which distinguishes five large and six smaller tourism resorts in Lapland with a tourist zone around them or connecting them (Figure 12.1). What is interesting from the point of view of northwest Lapland is that in these planning documents only three places are recognised as tourism resorts and two of them, Pallas and Olos in Muonio, have only one hotel with few or no other services. At the same time, around Muonio, especially around Pallas fells, there exist several small and medium-sized tourism enterprises within villages and places that are not recognised as resorts but, when it comes to service structure, do not differ much of those recognised ones.

In 1970, professor of geography, Uuno Varjo, forecasted that tourism or other services would not become a remarkable source of income in Finnish Lapland. However, he thought that the northwest municipalities of Muonio and Enontekiö with their long touristic histories might benefit from tourism more than other nearby municipalities. The past 30 years have proven that the situation is the other way around. Tourism has grown significantly in the nearby municipalities of Kittilä and Kolari with the big tourist resorts of Levi and Ylläs (Kauppila 2005). Nevertheless, it has also grown at a smaller scale in Enontekiö and Muonio, with Muonio in particular attracting foreign tourists. In any case, over recent decades tourism in northwest Lapland has grown rapidly and the whole area is now among the most important tourism regions outside the capital area in Finland with its two large tourist resorts and several smaller ones. Tourism growth in the Levi and Ylläs resorts has followed the common ideals of tourism destination development with highly concentrated services and massive infrastructure. Their development also follows the guidelines provided by the Finnish Tourism Strategy 2020 (Ministry of Trade and Industry 2006).

Tourism in the municipality of Muonio is somewhat different by nature compared with its neighbouring municipalities in the south. With the internationalisation of tourism, the need for activity programs and services has become more important, and entrepreneurs in Muonio have been more successful at meeting that need. In Muonio, the proportion of international tourists is higher than in any other destination in Finland. Tourism is wilderness- and nature-based, which is not the case in many other destinations in Lapland. Long distance hikes and safaris are more common in the Pallas region, and remote areas are used for different activities (tourist entrepreneur, interview 25.1.2011). Tourists in Muonio also consist of a large variety of groups from domestic, tradition-oriented skiing tourists, hikers and cottagers to foreign cross-country skiing training groups, adventure tourists, incentive groups and tourist-like car testers.

In addition, tourism employs relatively more people in Muonio than in other municipalities: 84% of employees of Muonio work in the service sector and much of this employment is related to tourism. Further, the proportion of foreign residents is bigger than in other municipalities in Lapland, although it is still relatively low (2.7% of the population). However, foreigners in Muonio, contrary to those in many other municipalities, have migrated there mostly because of tourism (interview with a trade promoter, 10.2.2012). What is also notable is that Muonio has two strong families that have created successful and nationally and internationally remarkable tourism enterprises. These issues, however, are not present in planning documents, as most tourism statistics concern the Fell Lapland sub-region and they are dominated by the big resorts of neighbouring municipalities.

Theoretical Background: Tourism Destination Development Paths

Tourism destination development is a widely discussed issue in tourism studies. The concept of tourism destination is also geographically wide-ranging from whole countries to small places. Some destinations, such as the Alps or Lapland, are somewhat artificially divided by nation states, although for consumers they seem to be parts of the same product (Buhalis 2000). The most important usually identified elements of a destination or resort are attractions, infrastructure and services, such as accommodation and transport (Butler 1980; Tinsley and Lynch 2001; Kauppila 2004; Bohlin and Elbe 2007; Saraniemi and Kylänen 2012). From a sociological perspective, social practices are emphasised (e.g. Framke 2002). Several classifications and categorisations from material presence to cognitive experience have been used in order to deconstruct and analyse the destination concept (e.g. Lew 1987; Framke 2002; Saraniemi and Kylänen 2012).

However, as Tinsley and Lynch (2001) stated, the listed destination features tend to be static and seen primarily from geographical and physical perspectives. From a more social and dynamic perspective, Frisk (1999, cited by Tinsley and Lynch 2001) recommended that a destination be interpreted as a place or region where some sort of 'touristic society' inside the 'ordinary' community exists. Socially and culturally, the destinations and other spaces of tourism are open to signification and to struggles over their representations and social meanings. In the end, space as a social category matters only through the medium of socio-cultural practices and structures manifested in certain spaces and through the different discourses of the transformation process (Saarinen 2004, 166).

In any case, what makes a 'destination' different from just a 'place' is that it is a stage of tourism-related social practices and is, as such, subjugated to specific modes of planning and public discussion. Thus, there are discourses that dominate when dealing with places that are mainly used for recreation and tourism. As Saraniemi and Kylänen (2012) noted, traditional models narrow the complex nature of destinations into strategic tourism decision making. What seems to be often forgotten is that one universal destination development does not exist but, as Bohlin and Elbe (2007, 20) stated, development can happen in different ways and it can be either planned or spontaneous. If we want to take socio-cultural practices seriously, we should speak more about destination cultures than destination development.

Destination development is, however, often seen as a result of purposeful planning by tourist organisations, policies and authorities (Bohlin and Elbe 2007). For example, Butler's model of a tourism area's life cycle has been appraised for offering a universal model that tourism studies otherwise lack (Hall 2006). The problem with it is, nevertheless, that in planning it has been seen as applicable to all destinations, although Butler himself emphasises that variation is nowadays wide and a unidirectional linear model is relatively unlikely to provide an accurate prediction of the future of a complex product subject to rapid change and great competition (Butler 2009, 348; Cole 2009).

Richards (2011) emphasised the role of tourists' and locals' activities in destination development and the creativity that arises from the interplay between consumers and producers (see also Saraniemi and Kylänen 2012). One example of this is the wide and important phenomenon of second home tourism that forms the backbone for low season tourism in many resorts in Lapland and creates a kind of tourist that is more attached to the place than just occasional visitors (cf. Tuulentie 2007).

When discussing the developmental paths of Pallas, it is important to remember the fairly long touristic history in the region. According to the idea of path dependence, history is not something that only precedes the present; instead, it makes some future developmental paths dominant, while others become marginal or impossible. Theoretically, for the idea of path dependency, it is important that, first, only early events in a sequence matter; second, these early events are contingent; and third, later events are inertial (Howlett and Rayner 2006; Mahoney 2000). However, there are many unresolved questions in relation to the concept of path dependency (cf. Martin and Sunley 2006). Nevertheless, we are here not trying to dig deeply into the concept of path dependence itself but to point out the outcomes of such turning points as the building of the hotel and the establishment of the national park, and their consequences for the future development of Pallas. Through the concept of path dependence, we aim to explain why the ideas of tourism development do not update with innovations and best practices (cf. Peltonen 2004). Being at the right place at a right time is more important than being the best. Inertia is also an important issue regarding path dependence in the Pallas case: a hotel building itself is not disposable, but once it has been built, it is a material fact that is intertwined with many kinds of social and cultural issues.

However, history does not determine development. There is always room for path creativity, which means that processes are not predetermined but also contingent and emergent. Moreover, it is better to talk about varying degrees of probability and of stronger and weaker tendencies of development that are inherent in the path dependence of social processes (Bramwell and Meyer 2007; Scott 2011). Further, such 'cross-path effects' (Schneiberg 2007; Martin 2010) exist in whichever developments in one institutional path influence those in another institutional path. In the case of Pallas, this means that the institutional structures of nature conservation, i.e. the national park, and those of tourism development, which co-exist in the Pallas area, have to adjust to each other.

In the case of Pallas, the focal features of the canonical path dependence model (Martin 2010) can also be described as follows. Firstly, the 'historical accident' was the discovery of Pallas as the best place for Finnish alpine skiing and thus accommodation was built in the place. Almost at the same time, the national park was established. Secondly, in the 'early path creation' phase, tourism started to grow in the region, and nearby villages also began to provide tourism services. For example, because of the lack of roads, transportation had to be arranged by local reindeer herders with their reindeer. Thirdly, what can be called 'path dependent lock-in' is the combination of the image of Pallas as a proper resort and the restrictions because of the national park's location. 'Path delocking' is outlined here based on the interviews with locals and experts.

Results

Path 1: Strong, independent resort in the context of nature conservation

The vision of Pallas as a strong, independent resort has prevailed since the 1970s as a kind of dominant path, which has been promoted by local and regional developers as well as entrepreneurs. This dominant idea has, however, not resulted in desired outcomes, which has meant that the hotel has remained physically almost unchanged.

In the past 50 years, there have been several attempts to renovate the hotel and to enlarge tourist services in Pallas. In 1967, an industrial committee appointed by the state suggested that in addition to one existing ski lift, another should be built. In addition, a cafe should be built next to the ski lift. The proposal also included more mid-range accommodation outside the national park. These actions were mainly carried out. After that, both the municipality of Kittilä (1974) and the municipality of Muonio (2004) suggested that the hotel should be enlarged and renovated but these plans did not lead to desired actions. Only a caravan site was added after the proposal in 1974. In the 1970s, plans were made for 1180 beds instead of the existing 130 beds, and in the 2000s the plan included 500 beds. The Government Bill 489/2010 regarding the use of the national park stated that the maximum number of hotel beds was 320 and, in relation to the hotel enlargement, that the caravan site should be removed. In addition, the enlargement of the visitor centre was connected to the development of the hotel.

Thus, the fact that hotel Pallas is situated inside the national park has prevented all enlargement suggestions except the last one in 2008, which was formulated as a governmental bill in 2010 (Hildén et al. 2009). However, the decision to reduce the allowed number of beds from 500 to 320 did not satisfy the entrepreneur, and the actual renovation did not start until the beginning of 2012. The administration of the Pallas area has also been a contested issue between municipalities: the government decided in 2003 (HE 566/2003) to transfer Pallas from the municipality of Kittilä to the municipality of Muonio. The initiative came from the owner of hotel Pallas from Muonio. The basic idea was that the municipality of Muonio would put more emphasis on the development of the place than Kittilä who already had a big ski resort in Levi.

Although Butler's (1980) model of a tourist area's life cycle was created for theoretical purposes, it has influenced the concrete development plans of Pallas. This was evident in the plans of Kittilä in the 1970s and in the first draft of the government bill in 2008, which was mainly based on the ideas of an innovation competition. The main guidelines in the proposal for the development of a successful resort were that the threat or reality of the decline stage with an emphasis on day visitors should be turned into a rejuvenation stage with longer stays and thus more profit. In order to change the direction of development, different kinds of accommodation designs both in price and in size were recommended, and it was regarded that turnover should be big enough to able the resort to offer diverse services. Thus, the basic idea was to develop Pallas into a strong independent resort that had a network of smaller service providers and other entrepreneurs in the nearby villages (Jaakko Pöyry Infra 2005).

The developmental path of Pallas is, however, shaped by the existence of the national park and the related institutional structures. In a national-level public debate, this view of developing Pallas as a 'normal' – meaning expanding – tourist resort was put against the basic idea of the national park, whose main purpose is to protect nature. It was stated that there are other ways of enhancing tourism in Lapland than expanding the infrastructure according to the government bill (e.g. Save Pallas-petition 2009).

Thus, the locally and regionally – but also nationally – dominant ideal of tourism development collided with the ideal of nature conservation. Until now, it seems as though the compromise will not satisfy any party, and the dilemma between conservation and tourism development is in this context unresolved (see also Rytteri and Puhakka 2012) and a kind of lock-in situation exists.

Path 2: Loose tourism agglomeration

What, then, are the more marginal paths that have been taking shape alongside the institutionally dominant but contested one? Instead of a strong independent resort model, there have been suggestions that the area should be seen as a network of entrepreneurs located in and around Pallas fells itself, which operate on an equal basis (Staffans et al. 2005).

This view also has its antecedents. In the Finnish tourism literature, Pallas has not always been regarded as a proper resort. Vuoristo and Vesterinen (2001), in their textbook, stated that only Ylläs and Levi are proper resorts in western Lapland. The area of Pallas is what they describe as a 'loose agglomeration of tourism services' that consists of the nearby villages and the neighbouring winter sports centre of Olos. This definition can be interpreted as referring to a different kind of resort that does not have to expand by itself, but which can be one part of a network. This network idea was also discussed in the winning proposal of the innovation competition (Jaakko Pöyry Infra 2005). It was called a 'regional model' but it was turned down in the report on the basis that day visitors would cause crowding without investing in the Pallas resort and that Butler's life cycle model sees a shift to a place for weekend or day trips as unsustainable. However, the 'regional model' saw Pallas only in relation to the big ski resorts, such as Levi and Ylläs, and a smaller one, Olos (Jaakko Pöyry Infra 2005, 6). The nearby villages were left out of this discussion, although it was emphasised that a strong independent Pallas resort would benefit the entrepreneurs in nearby villages as well (Jaakko Pöyry Infra 2005, 17).

The idea of a strong independent Pallas does not support the way the developmental paths of tourism in Muonio are seen from the local authorities' perspective. A trade promoter of Muonio (interview 10.2.2012) stated that the existing, more scattered tourism development has been a strategic choice and a different model compared to the big resorts of Levi and Ylläs, although he admitted that sometimes it has been more an accident than an intention. However, according to the trade promoter, there is always a push that Muonio should also have a big resort, and this pressure comes not only from the local level but also from national and regional strategic planning initiatives.

The municipality made a proposal for a mountain and lake district around Pallas. The alternative scenarios included the idea of concentrating either on domestic tourism with an increasing amount of second homes or on international tourism with more compact resorts. The trade promoter of Muonio interpreted these plans as targeting Pallas being in the centre but services being provided by the nearby villages.

From Lock-in to More Creative Solutions

The situation with Pallas development after the government bill in 2010 has been static. The entrepreneur who owns the hotel has expressed his dissatisfaction with the decision that hotel Pallas can have only 320 beds (Yle.fi 2011). He runs the hotel chain Lapland Hotels, which is among the 10 biggest hotel chains in Finland, and does not have to focus on Pallas just now. According to Lapland Hotels' CEO, the company is currently (2012) focusing on developing its Ylläs units and there probably will be no change in Pallas hotel's situation in many years. It sees the capacity of 320 beds as too small to ensure sufficient turnover to cover the investments that the enlargement of the hotel requires (interview with the director of Lapland Hotels, 13.2.2012).

If we take seriously Tinsley and Lynch's (2001) definition of destination as a place or region where some sort of 'touristic society' exists inside the 'ordinary community', then we have to direct attention to the surrounding villages of Pallas, as Pallas mountain itself does not have any permanent residents. These villages have as long traditions in tourism as the mountain resort itself, and thus a touristic society has existed for a long time. The villages have a lot of active second homeowners and foreign long-term visitors because of the landscape of Pallas, and in addition to numerous small tourism enterprises, most of the reindeer herders and traders of other more traditional livelihoods have tourism as their secondary occupations (Suunnittelukeskus Oy, no date). Another feature typical of Muonio, compared with some bigger resorts, is that the entrepreneurs, both big and small, are locals (trade promoter, interview 10.2.2012).

According to the group discussions, the ownership of Pallas hotel and the existence of the national park seem to be the major factors influencing touristic development in the Pallas area. The services of the mountain resort are dependent on a single entrepreneur's actions, and some locals have contemplated whether the development path of the area would have been different if there had have been more entrepreneurs in the core area offering more diverse services such as accommodation of different standards. The establishment of the national park is usually seen in a positive light as it brings tourists to the area. However, the limitations that the national park sets, for example on land use, are also seen as hindering the development of the area. In addition, the relationships between locals and park management are regarded as dependent on the personal features of the manager at the time.

Opinions on the suitable future development paths of Pallas fells and its surroundings vary among local people. According to the group discussions, the opinion shared by all is that something has to be done about the current situation. The biggest problems in the actual Pallas resort are the bad condition of the hotel and its short seasons. This deteriorates the image of the whole area and gives the place a drowsy feeling. No services are available in the core area when the hotel is closed except for the national park visitor centre, whose services are limited to an exhibition, some souvenirs and toilets. In general, because of its varying and entrepreneur-dependent opening hours, seasonality is seen as a growing problem compared with in the past. It has been claimed that the hotel is so devoted to certain foreign customer segments that the opening hours of the restaurant or hotel do not serve individual tourists, occasional visitors or locals.

The problems in the wider Pallas area include the poor condition of the hiking trails in the national park, which complicates the touristic development of the surrounding villages. Many interviewees see that these villages would have a lot to provide tourists that could not be offered in the core area of Pallas but the lack of infrastructure or its current condition hinders the development of these areas. In addition, the strict limitations of land use are regarded as a barrier to touristic development as well as to locals' everyday lives. Therefore, local villagers seem to support the idea of developing the surrounding villages to offer touristic services but the improvement of the hotel area is also seen to be crucial, as it acts as 'the heart of the area' and its development would benefit other tourism entrepreneurs in the surrounding area.

The possibilities of developing the hotel and its immediate surroundings are described as 'rather restricted' since the intense growth in the amount of tourists in the area can be seen as a threat to the surrounding environment, to traditional nature-based livelihoods and to the amenity values of the area. The important thing is to preserve the 'spirit of Pallas', such as its characteristics of being close to nature, calm, clean, secure and non-elitist. The area should be developed but in a way that it respects traditions, suits the national park environment and offers year-round services in cooperation with the other smaller entrepreneurs around the fells. The key factor to successful development is seen to be cooperation between different actors in the area, including local villagers, entrepreneurs, the municipality of Muonio and the Finnish state-owned forest administrator Metsähallitus. A development plan covering the whole region is regarded as important to coordinate with all these actors.

Although cooperation between these actors is considered to be crucial, it has not always been easy. According to the superintendent of the Pallas-Ylläs national park, Metsähallitus has improved its cooperation with local villages and entrepreneurs, and he perceives that cooperation will enhance and increase even more in the future. One form of cooperation is making agreements with local entrepreneurs in order to allow enterprises to utilise the park logo and to encourage them to follow and promote the principles of sustainable tourism (interview with the superintendent, 9.2.2012). In the nearby villages of Pallas, however, cooperation with Metsähallitus is not always seen as very productive, as some villages felt that the maintenance of routes and signs was not given enough support, which affected tourism development. Cooperation between villages was considered to be one solution to this problem, as villages could apply for financing from the EU for development projects. Cooperation with the entrepreneur of hotel Pallas was also perceived as complicated, as it seemed to be difficult to get him to negotiate with the other actors. Regardless of the difficulties in current cooperation, the hotel was considered to have an important role in the future development of the area.

Future Paths?

The conventional developmental path of Pallas can be described as consisting of three actors: 1) the national park, which puts limitations on and preconditions for the use of the environment and has the decision-making power over changes in the visitor centre, and which is managed by the state. Thus, it is also the main representative of national interests in the area; 2) the private entrepreneur who owns the long-standing hotel building in Pallas and 3) the small-scale tourism in surrounding villages, which suffers from a lack of capital and the aging of entrepreneurs. These parties create among villagers the feeling that everything runs along the same path and nothing develops. More path-creative solutions to development seem to be needed instead of following the guidelines of tourism strategies and concentrating on single-resort development.

However, during the discussions, multiple new ideas concerning development also came up. They varied from demolishing the hotel to creating a large fells resort with many services. One suggestion was to enhance the role of the visitor centre by renovating and expanding it and creating additional services within it such as a cafe and hostel-style accommodation for hikers. Another suggestion was to build the hotel outside the actual current tourist resort to allow it to be larger than the government bill permits. The better use of accommodation services in nearby villages was also suggested, which would require improving the transportation system in the area as well as the path network between Pallas fells and the villages. Overall, it was seen as important that different actors in the area would decide together what a suitable future development path for the area might be and not try to copy the development model from another destination.

So far, the common experience was that locals' voices had not been heard in planning discussions and that the interests of the national park administration and local residents were more or less conflictual. To some extent, this also applies to the relationships between the hotel owner and other locals, although this seems to be mainly a problem of communication.

Conclusion

Contrary to many other rural or peripheral regions in Nordic countries, northwest Lapland in Finland has a long history of tourism and development has not been tied only to traditional rural livelihoods. The combination of nature conservation and tourism has created a specific path for development: on one hand, the institutional structures of conservation have restricted the building of tourism infrastructure, while on the other hand, tourism development strategies and practices have been 'locked' into the idea of the development of large-scale resorts.

The discussion around Pallas makes two things visible: the arbitrariness of defining a tourist destination and the social and physical structures that create at least some kind of path dependence. In the case of Pallas, the critical turning points have been, in addition to its 'discovery' as a tourist destination, the establishment of the national park and thus the strong bonds to the state administration. Further, the physical existence of hotel buildings demand some kind of continuity, as their dismantling seems to be too radical a solution for all parties.

Likewise, a path-dependent feature is that of the dominant idea of destination development, which demands that a resort has to grow in order to be profitable and thus economically sustainable. No other options are relevant from the perspectives of national and regional tourism strategies or entrepreneurs. However, from a wider national point of view, tourism development would threaten the high amenity and conservation values of the place.

On these bases, the role of Pallas in the development of tourism in the area could go far beyond the standard paths of single resort-based thinking, and path-creative solutions could be brought forth. It is obvious that whatever the future decisions are, Pallas is facing a kind of turning point in its developmental path. It has become clear that the old paths have to be followed when it comes to conservation and its infrastructure and other land-use issues, but from the perspective of the vitality of communities and different modes of tourism, new – and more creative – paths are also needed. However, the Pallas case also challenges, at least to some extent, the big resort-driven thinking of tourism development as the only solution.

References

Bohlin, M. and Elbe, J. (2007), 'Utveckling av turistdestinationer – en introduktion', in Bohlin and Elbe (eds).
Bohlin, M. and Elbe, J. (eds) (2007), *Utveckla turistdestinationer. Ett svenskt perspektiv* (Uppsala: Uppsala Publishing House).
Bramwell, B. and Meyer, D. (2007), 'Power and Tourism Policy. Relations in Transition', *Annals of Tourism Research* 34:3, 766–788.
Buhalis, D. (2000), 'Marketing the Competitive Destination of the Future', *Tourism Management* 21:1, 97–116.
Butler, R. (1980), 'The Concept of a Tourist Area Cycle of Evolution: Implications for Management of Resources', *The Canadian Geographer* 24:1, 5–12.
Butler, R.W. (ed) (2006), *The Tourism Area Life Cycle, Vol. 1 Applications and Modifications* (Clevedon: Channel View Publications).
Butler, R. (2009), 'Tourism in the Future: Cycles, Waves or Wheels?', *Futures* 41, 346–352.
Cole, S. (2009), 'A Logistic Tourism Model. Resort Cycles, Globalization, and Chaos ', *Annals of Tourism Research* 36:4, 689–714.
Framke, W. (2002), 'The Destination as a Concept: A Discussion of the Business-Related Perspective versus the Socio-Cultural Approach in Tourism Theory', *Scandinavian Journal of Hospitality and Tourism* 2:2, 92–108.

Hakkarainen, M. and Tuulentie, S. (2008), 'Tourism's Role in Rural Development of Finnish Lapland: Interpreting National and Regional Strategy Documents', *Fennia* 186:1, 3–13.

Hall, C.M. (2006), 'Introduction', in Butler (ed).

Hildén, M., Kallio, T., Norokorpi, Y., Sulkava, P., Tyrväinen, L., Tuulentie, S., Sievänen, T. and Helle, T. (2009), *Pallas-Yllästunturin kansallispuistoa koskevan lakimuutoksen ympäristöarviointi.* Unpublished report. Suomen ympäristökeskus (Rovaniemi: Metsähallitus, Metsäntutkimuslaitos).

Howlett, M. and Rayner, J. (2006), 'Understanding the Historical Turn in the Policy Sciences: A Critique of Stochastic, Narrative, Path Dependency and Process-Sequencing Models of Policy-Making over Time', *Policy Sciences* 39, 1–18.

Jaakko Pöyry Infra (2005), *Pallaksen matkailukeskuksen toiminnallinen ideointi.* Unpublished report by the Jaakko Pöyry consulting firm (Muonio: Municipality of Muonio).

Kari, K. (1978), *Haltin valloitus* (Helsinki: Kisakalliosäätiö).

Kauppila, P. (2004), *Matkailukeskusten kehitysprosessi ja rooli aluekehityksessä paikallistasolla: esimerkkeinä Levi, Ruka, Saariselkä ja Ylläs.* Publications of the Geographichal Society of Northern Finland and the Department of Geography (Oulu: University of Oulu).

Kauppila, P. (2005), 'Ennustaminen on vaikeaa', *Kuukkeli magazine.* Available: <http://www.kuukkeli.com/4juttu1205.html> (home page), accessed 15 May 2013.

Lapinkävijät (no date), Available: http://lapinkavijat.rovaniemi.fi/vaeltaja/vaeltaja. html (home page), accessed 15 May 2013.

Lew, A.A. (1987), 'A Framework of Tourist Attraction Research', *Annals of Tourism Research* 14:4, 553–575.

Mahoney, J. (2000), 'Path Dependence in Historical Sociology', *Theory and Society* 29:4, 507–548.

Martin, R. (2010), 'Roepke Lecture in Economic Geography – Rethinking Regional Path Dependence: Beyond Lock-in to Evolution', *Economic Geography* 86:1, 1–27.

Martin, R. and Sunley, P. (2006), *Path Dependence and Regional Economic Evolution.* Paper prepared for the European Science Foundation Exploratory Workshop on Evolutionary Economic Geography, St Catherine's College, Cambridge, 3–5 April, 2006.

Metsähallitus (2011), *Käyntimäärät kansallispuistoittain 2010.* Available: <http://www.metsa.fi/sivustot/metsa/fi/Eraasiatjaretkeily/Asiakastieto/Kayntimaarat/Kansallispuistoittain/Sivut/kayntimaaratkansallispuistoittain2010.aspx> (home page), accessed 15 May 2013.

Ministry of the Environment (2008), *Draft for the Government Bill 12.3.2008* (Helsinki: Ministry of the Environment).

Ministry of Trade and Industry (2006), *Finland's Tourism Strategy to 2020.* 21/2006. (Helsinki: Ministry of Trade and Industry). Available: <http://www.mek.fi/w5/meken/index.nsf/(images)/Finlands_Tourism_Strategy_to_2020/$File/

Finlands_Tourism_Strategy_to_2020.pdf> (home page), accessed 15 May 2013.

Nyman, K. (2004), *Pallaksen kehittäminen*. Unpublished development plan (Muonio: Municipality of Muonio).

Peltonen, L. (2004), 'Paikallisen hallintatavan polkuriippuvuus – Tampereen aseveliakselin tie rintamalta regiimiksi', *Yhdyskuntasuunnittelu* 42:3–4, 30–50.

Regional Council of Lapland (2009), *Lappi – Pohjoisen luova menestyjä*. Lapin maakuntasuunnitelma 2030. Regional development plan (Rovaniemi: Regional Council of Lapland).

Regional Council of Lapland (2011), *Lapland Tourism Strategy 2011–2014*. A draft. (Rovaniemi: Regional Council of Lapland) Available: <http://lapinliitto. tjhosting.com/kokous/2011126-6-4523.PDF> (home page), accessed 15 May 2013.

Richards, G. (2011), 'Creativity and Tourism. A State of the Art', *Annals of Tourism Research*, 38:4, 1225–1253.

Rytteri, T. and Puhakka, R. (2012), 'The Art of Neoliberalizing Park Management: Commodification, Politics and Hotel Construction in Pallas-Yllästunturi National Park, Finland', *Geografiska Annaler: Series B, Human Geography* 94:3, 255–268.

Saarinen, J. (2004), '"Destinations in Change'. The Transformation Process of Tourist Destinations', *Tourist Studies* 4:2, 161–179.

Saraniemi, S. and Kylänen, M. (2012), 'Problematizing the Concept of Tourism Destination: An Analysis of Different Theoretical Approaches', *Journal of Travel Research* 50:2, 133–143.

Save Pallas-petition (2009), Available: <http://www.adressit.com/pelastapallas> (home page), accessed 2 October 2012.

Schneiberg, M. (2007), 'What's on the Path? Path Dependence, Organizational Diversity and the Problem of Institutional Change in the US Economy, 1900–1950', *Socio-Economic Review* 5:1, 47–80.

Scott, J. (2011), *Conceptualising the Social World. Principles of Sociological Analysis* (Cambridge: Cambridge University Press).

Sippola, A.-L. and Rauhala, J.-P. (1992), *Acerbin keinosta Jerisjärven tielle. Pallas-Ounastunturin kansallispuiston historiaa*. Metla research reports, 410 (Rovaniemi: Metla).

Staffans, A., Paloniemi, A., Verhe, I., Ruti, M., Ronkainen, H. and Heliövaara, K. (2005), *Parasta Pallakselle*. Unpublished proposal for the development of Pallas destination (Muonio: Municipality of Muonio).

Suunnittelukeskus Oy (no date), *Veturi. Muonion vesi- ja tunturialueitten kehittämissuunnitelma ja yleiskaavat*. [Development plan and general plan for the water and fell areas of Muonio] (Muonio: Municipality of Muonio).

Tinsley, R. and Lynch, P. (2001), 'Small Tourism Business Networks and Destination Development', *Hospitality Management* 20, 367–378.

Tuulentie, S. (2007), 'Settled Tourists: Second Homes as a Part of Touristic Life Stories', *Scandinavian Journal of Hospitality and Tourism* 7:3, 281–300.

Vuoristo, K.-V. and Vesterinen, N. (2001), *Lumen ja suven maa. Suomen matkailumaantiede* (Helsinki: WSOY).

Yle.fi (2011), Pallaksen hotellia ei kannata laajentaa (Enlargement of Pallas hotel is not profitable, says Lapland Hotels concern director Pertti Yliniemi) <http://yle.fi/alueet/lappi/2011/04/quotpallaksen_hotellia_ei_kannata_ laajentaaquot_2496760.html> (home page), accessed 15 May 2013.

PART IV
Destinations as Politics

Chapter 13

Dynamic Development or Destined to Decline? The Case of Arctic Tourism Businesses and Local Labour Markets in Jokkmokk, Sweden

Dieter K. Müller and Patrick Brouder

Introduction

Tourism is often purported to be a solution to economic decline arising from restructuring of other sectors of the economy. This is particularly true for remote places perceived as lacking alternative options for development. Thus, tourism development in peripheral areas is often triggered as a result of crisis in other sectors (Baum 1999; Hall and Boyd 2005; Hall 2007; Jóhannesson and Huijbens 2010). Tourism development rarely, however, provides the solution it is supposed to – either by failing to deliver socioeconomic change (Hall 2007) or by simply replacing one mono-sectoral dependency (e.g., extractive industries) with another one (e.g., tourism) (Schmallegger and Carson 2010). Therefore, it is necessary to explore whether tourism offers dynamic development or is destined to decline in the long-term.

Tourism in the peripheral areas of Sweden has been shown to have a positive impact on the supply of local goods and services (Löffler 2007). This development has not gone unnoticed and, as a consequence, multiple initiatives have been supported to develop destinations in the Swedish periphery. In particular, the availability of European Union funding enabled previously less touristic places to invest into projects aimed at creating attractions and destination organisations. The apparent increase of tourism in importance to local labour markets has, however, been more dependent on changes in other sectors, reflecting only a relative development of tourism in the long term (Müller and Ulrich 2007). This relative growth has ended during the last decade with extractive industries experiencing resurgence since 2005 (Müller 2013a). What then is the future of tourism development in Arctic Sweden? Is its development to be stunted by its inability to compete for labour with the more lucrative extractive industries? Is the focus of planning and politics to move away from endogenous growth and back to the paternal industrialism of the past? Or is there a third way where tourism takes its place as a part of overall local development by complementing some other sectors (e.g., extractive industry) and contributing to others (e.g., creative arts and local service supply), ultimately attaining a more diversified peripheral economy (Noakes and Johnston 2009)?

Entrepreneurial activity and institutional efforts have led to tourism growth in numbers of new firms but does this help to sustain communities? Particularly, research has not tracked the local development indicators of tourism over the long-term. This chapter is an initial attempt to redress this imbalance, combining longitudinal quantitative measures of the tourism labour market in Jokkmokk as well as a qualitative assessment of tourism's place in the locale in order to assess tourism's impact at the destination level over the long-term. The analysis is based on microdata covering the local population and interviews with tourism stakeholders in the municipality of Jokkmokk, which is an example of a destination that has struggled to develop a viable tourism economy despite the presence of attractions such as national parks and nature reserves. The discussion highlights the importance of considering both 'hard' and 'soft' measures of tourism development as well as tourism's interrelationship with other sectors of the local economy.

An important consideration is: what are the goals of destination development? Is destination development about delivering products and services to visitors or is it about utilising tourism growth as a resource for community development?(Moscardo 2008)? The chapter highlights the success (or failure) of tourism to deliver positive labour market outcomes as well as community-based outcomes and places this in the context of changes in destination development. If destination development is about more than just servicing tourists then such outcomes need to be studied more systematically over time, as this chapter attempts to do.

Tourism Development as the Last Straw

Many long established northern communities are facing multiple challenges to their survival. In this context, tourism is often a last chance strategy for local development. The development of many northern European and Arctic areas is the result of a historical interest in the abundant resource base of forests, hydropower and minerals (Sörlin 1988). This also led to an increasing population and the development of communities (Håkansson 2000). However, industrialisation often resulted in single industry towns only and a desired diversification of peripheral economies seldom happened (Hayter 2000; Müller 2013a). Instead, research and development, as well as other knowledge-intensive stages of the production process, remained outside the area, causing a leakage of capital from the periphery. The result has been labeled a 'truncated economy', acknowledging the missed opportunities related to a great inflow of financial capital and people into the periphery (Gunton 2003, 69). This was, in part, due to the priorities of government and industries, which were mainly interested in effective staple production, while regional development ideas tended to be less persistent, at least as long as employment in the staple industries was guaranteed (Markey et al. 2006). Thus, resource communities showed little resilience to the impacts of market fluctuations and restructuring. In the case of Sweden, Westin (2006) argued that investments in welfare were seen as compensation to locals for the stagnating development in less attractive and remote areas.

However, globalisation and the economic crisis of the 1990s required new approaches for dealing with decline. In this context, tourism was often identified as a way to provide alternative development paths in the northern peripheries (Müller 2011; 2013a). Hence, multiple initiatives have been launched in order to develop such destinations' competitiveness on both domestic and international markets. The desired outcomes of such developments are usually formulated in terms of economic growth, including new employment opportunities, in order to counteract ongoing outmigration from many peripheral places. In Sweden, this ambition of sustaining regional economies permeates almost all policy areas. Hence, even the Environmental Protection Agency is expected to protect in order to contribute to sustainable growth (Müller 2013b).

This is particularly relevant since the major resource for tourism development in northern peripheries has been access to pristine nature (Hall and Boyd 2005; Hall 2007). In contrast to urban areas, where competition between different forms of land use usually implies that only few are able to exist at a time, the more spacious conditions in peripheral destinations allow for multiple land uses to co-exist but also to continue to compete. As a result of recurrent waves of deindustrialisation and decline, the value of these assets has been reconsidered at various times. Owing to recent globalisation and deregulation of markets, interest in utilising these resources for tourism development has been high on the agenda for peripheral communities and northern states across Europe and North America. For example, Pedersen and Viken (1996) see the development of a global playground for adventure seekers in the north of Norway, building on the social construction of a last European wilderness. Also, Saarinen (2005) and Sæþórsdóttir et al. (2011) reveal that the social construction of touristic wilderness played a crucial role in the process of profiling northern areas as tourist destinations.

A basic idea for this development path is the value creation and commodification resulting from protection (Mose and Weixlbaumer 2007). By protecting a piece of land or a heritage site, a marker is created identifying the object as special and thus worthwhile to see and to visit. The process thus converts otherwise non-productive land into a commodity and ultimately to an input into a tourism production system including linkages to other parts of the local economy. Multiple studies have addressed this issue. For example, Wall Reinius and Fredman (2007) argue that national park designation is a powerful marker able to attract tourists. World heritage status also indicates that an area is of interest to a global market and hence, the uniqueness of the destination as expressed by these designations is expected to overcome constraints and costs caused by distance to demand markets (Prideaux 2002). However, nature-based tourism development may occur in non-protected areas as well, for example, the Swedish Environmental Protection Agency promotes nature protection as a tool for leveraging such development. This creates local expectations towards the economic impacts of tourism in protected areas with potential impacts seen as compensation for other development restrictions.

The Swedish EU-membership in 1995 also entailed access to structural funds promoting endogenous growth by mobilising cooperation of local stakeholders. The resulting public-private partnerships are frequently found within tourism, most often organised as projects aiming at promoting local tourism development (Müller 2007). It can be argued that this situation has also entailed a willingness to embrace ideas of destination development rather early. Both destination development and engagement in EU-funded projects require cooperation of various stakeholders in order to promote a common goal. However, certainly EU-membership has not been the only reason to promote destinations as a tool for tourism development. Globalisation also meant a greater need for marketing and a desire to become visible on international demand markets. However, financing marketing campaigns required a pooling of resources and, moreover, the need to adapt to international standards regarding hospitality and services required educational activities (Zillinger 2007). Thus the establishment of destinations appeared to be a self-evident choice for creating a platform for organising and conducting these activities.

From a geographical perspective, tourism development initiatives in peripheral areas are not particularly well motivated endeavours. The distance decay effect usually favours destinations with good accessibility to major demand markets and hence Lundgren (1982) argues that there is a destination hierarchy which puts peripheral destinations at a disadvantage. Still, peripheral areas often attempt to profile tourism as a sector for growth and employment creation (Hall and Boyd 2005; Müller and Jansson 2007). And indeed, this development has left some noticeable traces. Müller and Ulrich (2007) demonstrate that northern labour markets are increasingly dependent on tourism employment, though this has been a result of the withdrawal of other industries, too.

Tourism development has proven to be a complicated endeavour. Despite an increasing interest in tourism in these areas (Hall and Saarinen 2010), development remains problematic and repeatedly fails to deliver the desired outcomes in terms of employment and economic growth (Hall 2007). For example, tourism supply often overestimates current demand and many destinations have indeed registered declining tourism figures since the early 1990s (Lundmark and Müller 2010; Müller 2011). Yet differences within the northern periphery remain undetected despite their obvious existence.

Despite the mixed experiences regarding tourism development, in development discourses tourism has continuously been promoted as a viable path for sustaining northern communities. Thus, in development discourses, positive experiences from northern tourism hotspots like Rovaniemi and Kiruna are extrapolated and applied to other locations and differences in accessibility and resource bases are neglected with success for neighboring destinations seemingly taken for granted.

As shown by Müller (2011) for the case of northern Sweden, destinations with local airports managed to increase visitor numbers.

Tourism as a single development path has recently been contested. A rejuvenated interest in northern resources, like forests and iron ore, has led to a situation where tourism is no longer promoted as the only option for development (Müller 2011). Of course, other industries promise year-round employment and high salaries, conditions that the tourism industry often fails to provide (Müller 2013c).

These development options threaten the image of a pristine wilderness used in the promotion of northern destinations. Sæþórsdóttir (2010) showed that tourists perceive development paths implying alternative exploitations of resources – in her case the construction of hydropower dams – as threats to the desired wilderness experience. Similarly, Lundmark and Stjernström (2009) reveal the problems of tourism development based on nature protection. Accordingly, it takes away alternative development opportunities for local communities by reserving land for the production of tourism and the preservation of flora and fauna only. Critique against the protection for tourism strategy is also articulated from within the tourism industry and from the local population. Strict preservation regimes imply serious constraints for tourism entrepreneurs and locals and hence, national park establishment is not always welcomed anymore (Sandell 2005; Müller 2013b). Although public authorities are increasingly aware of the need to include local stakeholders in the process of establishing new national parks, they sometimes simply fail to deliver and hence lack trustworthiness when negotiating with local communities (Zackrisson et al. 2006; Müller 2013b).

Hence, in summary it can be noted that northern areas have originally been developed owing to their richness in natural resources. As a consequence of deindustrialisation, tourism has been promoted as an alternative tool for regional development in northern areas. However, experiences with destination development varied considerably and thus, rejuvenated interest in northern resources increasingly challenges tourism development in the North. However, it is obvious that expectations and experiences towards destination development vary between different stakeholders. This is now demonstrated for the case of Jokkmokk, Sweden.

Study Area

Jokkmokk has a population of 5,000 and is the second largest municipality in Sweden with a total area of 21,000 km². Originally, the area was settled by the Sámi people who used it as a meeting place for trade and since 1605 there has been a crown-authorised winter market in the village. In the twentieth century, the area was developed for hydro-electricity and historically most employment has come from Vattenfall (the state-owned hydro-power company) and the government (primarily in healthcare, social work, and education). Reindeer husbandry has been a way of life for many Sámi in Jokkmokk and there is a history of exploitation and disempowerment of the local Sámi, not least in matters concerning hydro-power development in traditional reindeer-grazing areas (Green 2009). The municipality also includes the Laponia World Heritage Area, the only mixed world heritage site in the Nordic countries, which is protected for its natural beauty, created by the receding glaciers of the last ice age, and for its Sámi heritage. The area contains four national parks and a number of nature reserves (see Figure 13.1). Tourism in Jokkmokk is quite varied with a blend of domestic tourists engaging in summer hiking, snow-scooter riding, and attending the winter market, as well as a significant group of international tourists interested in both Sámi culture and nature. Only recently has tourism entrepreneurship been treated as a serious endeavour (Brouder and Eriksson 2013) and there is also potential for growth in cultural tourism now that the Laponia Area is under autonomous management (Brouder 2012).

However, tourism figures in Jokkmokk mirror the long-term decline in the north of Sweden over the last 20 years. Commercial accommodation figures dropped from 92,000 in 1990 to 75,000 in 2006 with foreign tourists representing 19% and 17.5% respectively. During the same period tourism changed dramatically – traditionally a summer destination the area increasingly profiles itself around winter products implying two seasons and improved preconditions for tourism businesses (Müller 2011). Further tourism development is being planned for in the municipality. In 2011, a new DMO – Destination Jokkmokk – was established to improve tourism development and increase the number of tourists to the municipality. It remains to be seen whether this new endeavour will result in long-term gain for the community or whether tourism in the municipality is destined to decline, particularly in light of the recent discovery of mineral deposits in the municipality with the possibility for a new wave of resource development impacting the village and the local tourism sector. Jokkmokk municipality has a growing culture of entrepreneurship, ranking highest in Norrbotten County for startups per capita in 2010 and 2011 (Nyföretagarcentrum 2011). It remains to be seen, however, whether this growth in startups is an indicator of a positive business climate – including tourism development – or whether it is a desperate combination of labour market policy support and individuals with few alternative employment options.

Method

This chapter uses a mixed methods approach to study the dynamics of local tourism development in Jokkmokk. The nature of a mixed methods approach opens the scope of the researchers' inquiry (Olsen and Morgan 2005) to the local nuances which are characteristic of rural communities. This chapter uses quantitative and qualitative data to examine local tourism. By incorporating both in the analysis, it is possible to come closer to understanding whether tourism development in Jokkmokk is dynamic and positive over time or whether it is destined to decline. Few studies incorporate both 'hard' and 'soft' perspectives while most scholars would acknowledge both are necessary to better understand destination development.

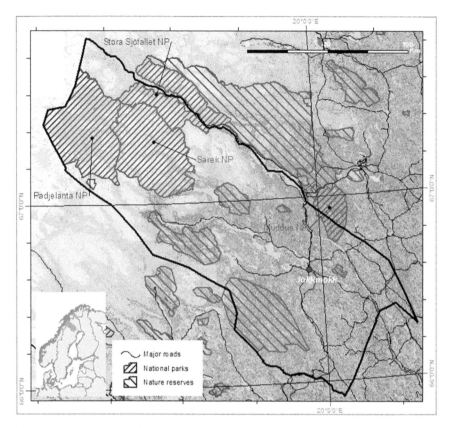

Figure 13.1   Protected nature in Jokkmokk municipality
Source: Dieter Müller 2013

The chapter uses longitudinal data from Statistics Sweden to examine the long-term development of the tourism labour market in Jokkmokk municipality. The data are drawn from ASTRID – a geo-referenced relational database from the Department of Geography and Economic History in Umeå University, Sweden. Standard Industrial Classification (SIC) codes at the 5 digit level help identify the local tourism labour force (see Table 13.1). The use of longitudinal data thus reveals the dynamics in the labour market over the long-term.

The quantitative data are complemented by qualitative interviews with local tourism stakeholders (six entrepreneurs and four institutional actors). The institutional actors come from local government authorities, some of whom are responsible for tourism and some of whom are responsible for other aspects of local economic development. The entrepreneurs represent a combination of nature-based tourism entrepreneurs and cultural entrepreneurs who have tourism as a part of their business. The interviewees were chosen for their long-term engagement with local development in Jokkmokk, both tourism and non-tourism development.

Table 13.1   Tourism SIC codes according to 2002 standard

Code	Key
55101	Hotels with restaurant except conference centres
55102	Conference centres with lodging
55103	Hotels and motels without restaurant
55210	Youth hostels etc.
55220	Camping sites including caravan sites
55230	Other short-stay lodging facilities
55300	Restaurants
55400	Bars
55521	Catering to the transport sector
61200	Inland water transports
62100	Scheduled air transports
62200	Non-scheduled air transports
63301	Activities of tour operators
63302	Activities of travel agencies
63303	Tourist assistance
92320	Operation of arts facilities
92330	Fair and amusement park activities
92340	Other entertainment activities
92520	Museum activities and preservation of historical sites and buildings
92530	Botanical and zoological gardens and nature reserves activities
92611	Operation of skiing facilities
92612	Operation of golf courses
92729	Various other recreational activities
52485	Retail sale of sports and leisure goods

Thus, they offer insights which complement the longitudinal, quantitative trends of tourism development in the municipality.

Semi-structured interviews were carried out with the interview guide focussing on three broad themes: economy, society, and institutions. Tourism was not the direct focus for institutional interviewees while it was more central for the tourism entrepreneurs. The rationale of the interview process was to place tourism development in the context of general local development. Post-interview analysis, however, focussed on tourism's changing place in Jokkmokk over the long-term, thus facilitating the particular analysis included in this chapter.

Results

From a regional economic perspective, tourism development in Jokkmokk has not been a success story despite the large number of protected areas and the establishment of the Laponia world heritage site. The development of registered commercial guest nights was in fact negative since 1990 (see Figure 13.2). Only 1995 was comparable to the level of 1990. Thereafter figures were around 20% lower. This is in sharp contrast to the development in Norrbotten County where a peak was registered in 1995. In the following years development stagnated but at a higher level. During the 2000s, figures were approximately on the same level as in 1995 again, reaching a new peak at the end of the covered time period in 2006.

The development in Norrbotten has to be seen against the background of two particular developments. First, the establishment of the Icehotel in Jukkasjärvi created a must-see attraction attracting tourists from all over the world to the northern town (Müller 2011). Moreover, the Icehotel marked the development of a new product category, namely winter tourism products, creating a second tourism season besides the earlier established summer season. The second development worth mentioning is the establishment of car-testing enterprises and facilities in and around Arvidsjaur municipality during the 2000s. Here, car and tire producers from all over the world test their equipment in winter conditions. This has created a seasonal demand for commercial overnight stays related to the car-testing industries. Originally being purely business tourism, it increasingly entails more leisure-oriented spin-offs offering driving on frozen lakes as a tourism product to individual travelers. However, despite these successful cases of tourism development in nearby municipalities, there seems to be no spillover effects on Jokkmokk. Neither did Jokkmokk as a destination copy the successful strategies of the other destinations in order to achieve a similar tourism development around winter tourism products.

Still, the role of tourism on the labour market in Jokkmokk should not be underrated. As in many northern municipalities, a majority of the labour is employed within the public sector catering not only for administration and maintenance, but also for education, child and elderly care. In 1990 a total of 184 persons were working in tourism-related industries. In 2007 the number has risen to 195, which is even more noteworthy considering the declining overall population figures in the municipality. Still a peak was reached in 1995 when almost 300 persons were employed within tourism. Hence despite a negative development of commercial overnight stays, employment figures in tourism increased. This has not only been as a result of the overall economic development in the region and municipality. Rather, it has been shown elsewhere (Müller 2013a) that the increase of tourism employment has been accompanied by a decline in other industries, most notably forestry, previously being a dominant force on the local labour market.

A comparison of different segments of the local labour market, namely tourism and staple industries – here defined as forestry and mining – indicates that tourism employment in Jokkmokk in 1990 did not match employment created

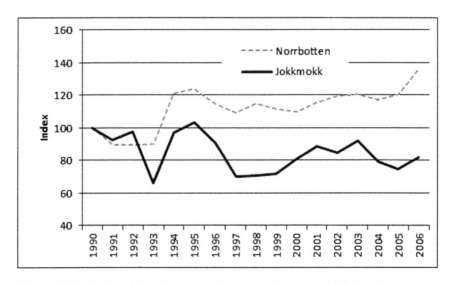

**Figure 13.2   Indexed development of commercial overnight stays in
Norrbotten county and Jokkmokk municipality, 1990–2006**

Source: Statistics Sweden, 2007 and Regionfakta 2012

in forestry. However, the economic crisis in Sweden, caused by an overheated property market and a consequent devaluation of the Swedish currency in 1992, hit staple industries in particular and led to a restructuring towards tourism and other service industries. Hence from 1993 onwards, tourism employment outweighed employment in forestry (see Figure 13.3). A recovery of the economy entailed, however, an increase of forestry employment and since 2005 the two sectors are of equal importance. Development elsewhere in Norrbotten mirrors the situation in Jokkmokk. In particular, Arvidsjaur, with a similar stronghold in forestry, shares Jokkmokk's development on the labour market. Mining in Kiruna and Gällivare meant that these municipalities never experienced a domination of tourism employment on the local labour market. Finally, the situation in Arjeplog demonstrates the impact of a mine closure leading to a radical restructuring of the local labour market.

Finally the quantitative analysis reveals an almost constant increase of firms within the tourism sector. In 1991 altogether 27 tourism firms could be found within the municipality. At the end of the study period 40 firms can be detected. Still there is significant turbulence in the sector and many firms have short duration only. Furthermore, growth can be mainly seen in micro firms with one to three workers only. The development of enterprises within tourism correlates only loosely with the development of commercial overnight stays. An increase can be noted, in particular, for the 2000s.

In summary, Jokkmokk did not share the positive development of commercial tourism overnight stays recorded elsewhere in Norrbotten. Despite the presence

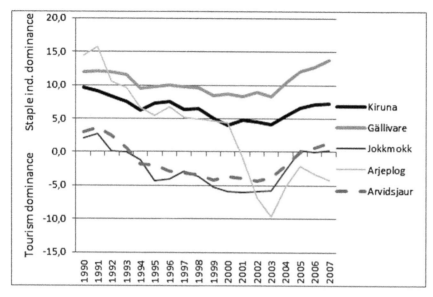

**Figure 13.3   Labour market development in tourism and staple
industries mining and forestry for selected municipalities in
Norrbotten, 1990–2007**

Source: The Astrid Database, Umeå University

of resources for nature-based tourism, and in particular the establishment of the
World Heritage area, tourism development did not take off, a fact that Müller
(2013c) explains is at least in part due to poor accessibility. Instead, tourism
struggles to maintain its position in the local economy.

The results from the interviews offer a different perspective on tourism
development in Jokkmokk. As a destination, Jokkmokk village is the main
beneficary of tourism in economic and social terms. However, it must be noted
that the Laponia area is both a distinct destination in its own right while also being
an important flagship attraction for tourism to the destination of Jokkmokk. Since
most commercial opportunities are realised in the village, it is there where the
interviews are focused but it is worth bearing in mind that Laponia is managed
as an independent destination and both Laponia and Jokkmokk are more than just
destinations for tourists – both have living societies using them for their lives
and livelihoods with tourism only being a part of the equation alongside forestry,
reindeer herding, etc.

Tourism is seen in a more positive light as time has passed with tourism
entrepreneurship being seen as a real and tenable occupation nowadays. One
interviewee stated: 'if you were in tourism a few years ago you were not really
considered a businessperson, now we are seen in a much better way' (entrepreneur
#8). However, alongside this acceptance of tourism entrepreneurship, there is
also a certain reticence among local authorities to take responsibility for tourism

development: 'the municipality tells the business people you must take care of this, tourism development, yourselves' (entrepreneur #4). This shows that the municipality does not see tourism development as their responsibility and implies that they see the (economic) value of tourism benefitting the entrepreneurs directly and it also reflects the pressure on municipal resources with little scope for investment in perceived marginal sectors.

Despite the municipality pulling back from responsibility for tourism development, tourism has a more noticeable, but slowly-developing role in community planning: 'it is coming more and more, it is slow but it is becoming more important' (institution #2). Tourism makes a number of obvious contributions locally, such as increasing the supply of services, with all interviewees acknowledging the fact that 'we have five restaurants and two rather big grocery stores and two hotels as well as a number of coffee shops in a municipality of 5,000' (institution #1). Thus, at the local level, tourism has palpable positive effects which may not be directly reflected in the labour market developments related to tourism. In a broader sense, the endogenous tourism development is a source of pride for the local community: 'there is value to have something like tourism that we developed on our own' (institution #4). Therefore, tourism development can have positive outcomes even if it is not growing rapidly. With development being left to the entrepreneurs, it is not so clear how that development should be driven. One interviewee thought: 'it would be good with one or two companies with more muscles' (institution #3). This is because small-scale tourism operations cannot leverage resources to develop the sector and many entrepreneurs are unable or unwilling to develop their own business: the entrepreneurs 'don't have the time and the money' (entrepreneur #5) and 'don't want to grow' (entrepreneur #1). Thus, tourism development remains challenging.

The local solution has been to establish a new DMO – Destination Jokkmokk – a company part-owned by the municipality and local entrepreneurs and run by the entrepreneurs through a managing director. The aim is to grow tourism in Jokkmokk in line with the national strategy for tourism (Svensk Turism 2010) and community development is seen more as an outcome of tourism growth but the focus remains on traditional metrics of development (overnight stays and visitor spending). The DMO also sees itself collaborating with neighbouring municipalities and the Laponia management to lead to broader development. It is too soon to assess its progress but surely its long-term success should be judged not only by its impact on the 'harder' measures of growth but also on how it facilitates 'softer' gains also.

Discussion and Conclusion

The case study certainly demonstrates that descriptions of reality, in this case the development of tourism and destinations, vary depending on the chosen method. A measurable downturn in tourism does not necessarily imply fading support for the sector and for attempts to develop destinations. This can be due to various reasons. As Müller (2013c) noted, tourism development may represent a local attempt at clutching at straws since the community is aware of lacking alternatives in other sectors. However, the development can even indicate that tourism benefits are perceived in terms of 'soft' impacts not measurable as employment or spending. Hence, the presence of service infrastructure is attributed to the presence of a tourism industry and locally the fact that many companies remain small is not seen as a failure but rather as a lifestyle choice. Thus the rationale of regional development policies which are usually focused on growth and employment creation is not totally accepted by local stakeholders. Instead community issues related to local quality of life are seen as worthy outcomes of development.

Ironically, this may be seen as a reaction to public development rhetoric regarding tourism. Since the often prescribed formula of nature protection for tourism development fails to deliver new employment, alternative benefits are realised and appreciated. Hence, the presence of a tourism industry and the awareness that tourists come to Jokkmokk to experience the North seems to be satisfactory for local stakeholders. To what extent this mirrors a general mistrust towards development strategies prescribed by national and regional politics has not been analysed in this study. However, evidence from other places in northern Sweden indicates that this may be the case (Müller 2013b). Still, not even Jokkmokk manages to totally disconnect from growth discourses exercised in policy and hence, new attempts are made to follow the beaten track, this time by establishing a new DMO. While the interviews were conducted before the DMO has had a chance to show its worth, the interviews showed a desire from both municipal and entrepreneurial stakeholders to have an empowered group of entrepreneurs and a more concerted effort to develop the destination. The DMO has set out on this track but the development of this institution will take time and is dependent on the co-evolution of the entrepreneurs and the DMO itself over time.

For a long time, tourism development, and more recently destination development, has been promoted as a tool for development in peripheral areas. However, often it has failed to deliver (Hall 2007). This may, however, be a question of perception and definition. For what is development in peripheral tourism? Certainly from a public point of view this is often about employment, entrepreneurship and growth. This study indicates that development and even destination development from a local perspective may mean different things – here local pride and quality of life are important contributions to community life. Hence, there is reason to interrogate definitions of development and acknowledge different perspectives on the role and impact of tourism. Obviously, local stakeholders do not seem to have the same expectations as regional and national stakeholders regarding economic development. Maintaining some level of economic development is thus an achievement in itself, a notion that may be understandable considering the long trends of negative development. Nevertheless, it seems obvious that different perspectives on development are interrelated and hence employment and business development cannot be ignored by local stakeholders either.

Thus, the community must decide what exactly its goals for destination development are. Is it to increase visitor numbers and expenditure resulting in more jobs in the tourism sector? Or is it to yield community benefits through modest tourism development? In conclusion, it seems that, notwithstanding the failure to deliver many new jobs, destination development in Jokkmokk is not necessarily destined to decline. The tenacity of tourism in the locale means it continues to offer something to a diversified local economy with meaningful community benefits. An important question for the new DMO remains: is the remit of the DMO to develop the tourism sector or to contribute to local development goals through tourism development? In this context it will be important to see how various stakeholders think and act regarding the destination. Is it a place for tourists only or does it include a community that is more than just one of the local attractions? The answers to these questions may well shape the future of not just the tourism sector but the community of Jokkmokk.

References

Almeida, P. and Kogut, B. (1999), 'Localization of Knowledge and the Mobility of Engineers in Regional Networks', *Management Science* 45:7, 905–917.

Baum, T. (1999), 'The Decline of the Traditional North Atlantic Fisheries and Tourism's Response: The Cases of Iceland and Newfoundland', *Current Issues in Tourism* 2:1, 68–81.

Beaudry, C. and Schiffauerova, A. (2009), 'Who's Right, Marshall or Jacobs? The Localization versus Urbanization Debate', *Research Policy* 38:2, 318–337.

Bohlin M. and Elbe J. (eds) (2007), *Utveckla turistdestinationer: ett svenskt perspektiv* (Uppsala: Uppsala Publishing House).

Brouder, P. (2012), 'Creative Outposts: Tourism's Place in Rural Innovation', *Tourism Planning & Development* 9:4, 383–396.

Brouder, P. and Eriksson, R.H. (2013), 'Staying Power: What Influences Micro-Firm Survival in Tourism?', *Tourism Geographies* 15:1, 124–143.

Buckley, R. (2004), 'The Effects of World Heritage Listing on Tourism to Australian National Parks', *Journal of Sustainable Tourism* 12:1, 70–84.

Butler, R.W. and Boyd, S.W. (eds) (2000), *Tourism and National Parks: Issues and Implications* (Chichester: Wiley).

Butler, R. et al. (eds) (1998), *Tourism and Recreation in Rural Areas* (Chichester: Wiley).

Carson, D. et al. (eds) (2011), *Demography at the Edge: Remote Human Populations in Developed Nations* (Aldershot: Ashgate).

Dearden, P. and Rollins, R. (eds) (2008), *Parks and Protected Areas in Canada: Planning and Management* (3rd ed, New York: Oxford University Press).

Eagles, P.F.J. and McCool, S.F. (eds) (2002), *Tourism in National Parks and Protected Areas: Planning and Management* (Wallingford: CABI).

Fleischer, A. and Felsenstein, D. (2000), 'Support for Rural Tourism: Does It Make a Difference?', *Annals of Tourism Research* 27:4, 1007–1024.

Fredman, P. and Yuan, M. (2011), 'Primary Economic Impacts at Three Spatial Levels: The Case of Fulufjället National Park, Sweden', *Scandinavian Journal of Hospitality and Tourism* 11:Suppl., 74–86.

Fredman, P., Hörnsten Friberg, L. and Emmelin, L. (2007), 'Increased Visitation from National Park Designation', *Current Issues in Tourism* 10:1, 87–95.

Gunton, T. (2003), 'Natural Resources and Regional Development: An Assessment of Dependency and Comparative Advantage Paradigms', *Economic Geography* 79:1, 67–94.

Grenier, A.A. and Müller, D.K. (eds) (2011), *Polar Tourism: A Tool for Regional Development* (Montreal: Presses de l'Université du Québec).

Hall, C.M. (2007), 'North-South Perspectives on Tourism, Regional Development and Peripheral Areas', in Müller and Jansson (eds).

Hall, C.M. and Boyd, S. (eds) (2005), *Nature-Based Tourism in Peripheral Areas: Development or Disaster?* (Clevedon: Channel View).

Hall, C.M. and Boyd, S. (2005), 'Nature-Based Tourism in Peripheral Areas: Introduction', in Hall and Boyd (eds).

Hall, C.M. and Saarinen, J. (eds) (2009), *Tourism and Change in Polar Regions: Climate, Environment and Experience* (Oxon, UK: Routledge).

Hall, C.M. and Saarinen, J. (2010), 'Polar Tourism: Definitions and Dimensions', *Scandinavian Journal of Hospitality and Tourism* 10:4, 448–467.

Håkansson, J. (2000), *Changing Population Distribution in Sweden: Long Term Trends and Contemporary Tendencies* (Umeå: Department of Social and Economic Geography).

Hammer, T. (2007), 'Protected Areas and Regional Development: Conflicts and Opportunities', in Mose (ed).

Haukeland, J.V., Daugstad, K. and Vistad, O.I. (2011), 'Harmony or Conflict? A Focus Group Study on Traditional Use and Tourism Development in and around Rondane and Jotunheimen National Parks in Norway', *Scandinavian Journal of Hospitality and Tourism* 11:Suppl., 13–37.

Hayter, R. (2000), 'Single Industry Resource Towns', in Sheppard and Barnes (eds).

Jenkins, J.M., Hall, C.M. and Troughton, M. (1998), 'The Restructuring of Rural Economies: Rural Tourism and Recreation as Government Response', in Butler et al. (eds).

Jóhannesson, G.T. and Huijbens, E.H. (2010), 'Tourism in Times of Crisis: Exploring the Discourse of Tourism Development in Iceland', *Current Issues in Tourism* 13:5, 419–434.

Kaltenborn, B.P., Qvenlid, M. and Nellemann, C. (2011), 'Local Governance of National Parks: The Perception of Tourism Operators in Dovre-Sunndalsfjella National Park, Norway', *Norsk Geografisk Tidsskrift* 65:2, 83–92.

Kasuhisa, I. et al. (eds) (2006), *Social Capital and Development Trends in Rural Areas. Vol 2* (Kyoto: Kyoto University).

Kauppila, P., Saarinen, J. and Leinonen, R. (2009), 'Sustainable Tourism Planning and Regional Development in Peripheries: A Nordic View', *Scandinavian Journal of Hospitality and Tourism* 9:4, 424–435.

Lemelin, R.H. and Johnston, M. (2008), 'Northern Protected Areas and Parks', in Dearden and Rollins (eds).

Löffler, G. (2007), 'The Impact of Tourism on the Local Supply Structure of Goods and Services in Peripheral Areas: The Example of Northern Sweden', in Müller and Jansson (eds).

Lundgren, J.O.J. (1982), 'The Tourist Frontier of Nouveau Quebec: Functions and Regional Linkages', *Tourist Review* 37:2, 10–16.

Lundmark, L. (2005), 'Economic Restructuring into Tourism in the Swedish Mountain Range', *Scandinavian Journal of Hospitality and Tourism* 5:1, 23–45.

—— (2006), 'Mobility, Migration and Seasonal Tourism Employment: Evidence from Swedish Mountain Municipalities', *Scandinavian Journal of Hospitality and Tourism* 6:3, 197–213.

Lundmark, L. and Müller, D.K. (2010), 'The Supply of Nature-Based Tourism Activities in Sweden', *Tourism* 58:4, 379–393.

Lundmark, L. and Stjernström, O. (2009), 'Environmental Protection: An Instrument for Regional Development? National Ambitions versus Local Realities in the Case of Tourism', *Scandinavian Journal of Hospitality and Tourism* 9:4, 387–405.

Lundmark, L.J.T., Fredman, P. and Sandell, K. (2010), 'National Parks and Protected Areas and the Role for Employment in Tourism and Forest Sectors: A Swedish Case', *Ecology and Society* 15:1, article 19.

Malmberg, A. (1996), 'Industrial Geography: Agglomeration and Local Milieu', *Progress in Human Geography* 20:3, 392–403.

Marsh, J. (2000), 'Tourism and National Parks in Polar Regions', in Butler and Boyd (eds).

Markey, S., Halseth, G. and Manson, D. (2006), 'The Struggle to Compete: From Comparative to Competitive Advantage in Northern British Columbia', *International Planning Studies* 11:1, 19–39.

Maskell, P. and Malmberg, A. (1999), 'Localised Learning and Industrial Competitiveness', *Cambridge Journal of Economics* 23:2, 167–185.

Mayer, M., Müller, M., Woltering, M., Arnegger, J. and Job, H. (2009), 'The Economic Impact of Tourism in Six German National Parks', *Landscape and Urban Planning* 97:2, 73–82.

Moisey, R.N. (2002), 'The Economics of Tourism in National Parks and Protected Areas', in Eagles and McCool (eds).

Moscardo, G. (2008), *Building Community Capacity for Tourism Development* (Wallingford: CABI).

Mose, I. (ed) (2007), *Protected Areas and Regional Development in Europe: Towards a New Model for the 21st Century* (Aldershot: Ashgate).

Mose, I. and Weixlbaumer, N. (2007), 'A New Paradigm for Protected Areas in Europe?', in Mose (ed).

Moss, L.A.G. (ed) (2006), *The Amenity Migrants: Seeking and Sustaining Mountains and their Cultures* (Wallingford: CABI).

Müller, D.K. (2006), 'Amenity Migration and Tourism Development in the Tärna Mountains, Sweden', in Moss (ed).

—— (2007), 'Planering för turistdestinationer', in Bohlin and Elbe (eds).

—— (2011), 'Tourism Development in Europe's "Last Wilderness": An Assessment of Nature-based Tourism in Swedish Lapland', in Grenier and Müller (eds).

—— (2013a), 'Hibernating Economic Decline? Tourism and Labor Market Change in Europe's Northern Periphery', in Visser and Ferreira (eds).

—— (2013b), 'National Parks for Tourism Development in Sub-arctic Areas – Curse or Blessing? The Case of a Proposed National Park in Northern Sweden', in Müller et al. (eds).

—— (2013c), 'National Parks, Protected Areas and Tourism Labor Markets in Arctic Sweden', *Journal of Sustainable Tourism* (Submitted).

Müller, D.K. and Jansson, B. (2007), 'The Difficult Business of Making Pleasure Peripheries Prosperous: Perspectives on Space, Place and Environment', in Müller and Jansson (eds).

Müller, D.K. and Jansson, B. (eds) (2007), *Tourism in Peripheries: Perspectives from the Far North and South* (Wallingford: CABI).

Müller, D.K. and Kuoljok Huuva, S. (2009), 'Limits to Sami Tourism Development: The Case of Jokkmokk, Sweden', *Journal of Ecotourism* 8:2, 115–127.

Müller, D.K. et al. (eds) (2013), *New Issues in Polar Tourism: Communities, Environments, Politics* (Dordrecht: Springer).

Müller, D.K. and Ulrich, P. (2007), 'Tourism Development and the Rural Labour Market in Sweden, 1960–1999', in Müller and Jansson (eds).

Noakes, J.L. and Johnston, M.E. (2009), 'Constraints and Opportunities in the Development of Diamond Tourism in Yellowknife, Northwest Territories', in Hall and Saarinen (eds).

Paniagua, A. (2002), 'Urban-rural Migration, Tourism Entrepreneurs and Rural Restructuring in Spain', *Tourism Geographies* 4:4, 349–371.

Pedersen, K. and Viken, A. (1996), 'From Sami Nomadism to Global Tourism', in Price (ed).

Price, M.F. (ed) (1996), *People and Tourism in Fragile Environments* (Chichester: Wiley).

Prideaux, B. (2002), 'Building Visitor Attractions in Peripheral Areas: Can Uniqueness Overcome Isolation to Produce Viability?', *International Journal of Tourism Research* 4:5, 379–389.

Puhakka, R., Sarkki, S., Cottrell, S.P. and Siikamäki, P. (2009), 'Local Discourse and International Initiatives: Sociocultural Sustainability of Tourism in Oulanka National Park, Finland', *Journal of Sustainable Tourism* 17:5, 529–549.

Quinn Patton, M. (2002), *Qualitative Research and Evaluation Methods* (London, UK: Sage).

Regionfakta (2012), Gästnätter mm månadsvis 2010 ff (hotell, stugbyar och vandrarhem): Norrbottens län. < http://www.regionfakta.com/Norrbottens-lan/Snabbstatistik/Gastnatter-pa-hotell-stugbyar-och-vandrarhem/ >.

Saarinen, J. (2003), 'The Regional Economic of Tourism in Northern Finland: The Socio-economic Implications of Recent Tourism Development and Future Possibilities for Regional Development', *Scandinavian Journal of Hospitality and Tourism* 3:2, 91–113.

—— (2005), 'Tourism in the Northern Wildernesses: Wilderness Discourses and the Development of Nature-based Tourism in Northern Finland', in Hall and Boyd (eds).

—— (2007), 'Tourism in Peripheries: The Role of Tourism in Regional Development in Northern Finland', in Müller and Jansson (eds).

Sandell, K. (2005), 'Access, Tourism and Democracy: A Conceptual Framework and the Non-establishment of a Proposed National Park in Sweden', *Scandinavian Journal of Hospitality and Tourism* 5:1, 63–75.

Sæþórsdóttir, A.D. (2010), 'Tourism Struggling as the Icelandic Wilderness is Developed', *Scandinavian Journal of Hospitality and Tourism* 10:3, 334–357.

Sæþórsdóttir, A.D., Hall, C.M. and Saarinen, J. (2011), 'Making Wilderness: Tourism and the History of the Wilderness Idea in Iceland', *Polar Geography* 34:4, 249–273.

Schmallegger, D. and Carson, D. (2010), 'Is Tourism just Another Staple? A New Perspective on Tourism in Remote Regions', *Current Issues in Tourism* 13:3, 201–221.

Schmallegger, D., Harwood, S., Cerveny, L. and Müller, D. (2011), 'Tourist Populations and Local Capital', in Carson et al. (eds).

Sheppard, E. and Barnes, T.J. (eds) (2000), *A Companion to Economic Geography* (Oxford: Blackwell).

Svensk Turism (2010), *Nationell strategi för svensk besöksnäring* (Stockholm: Strategi för svensk besöksnäring).

Sörlin, S. (1988), *Framtidslandet: Debatten om Norrland och naturresurserna under det industriella genombrottet* (Stockholm: Carlssons).

Visser, G. and Ferreira, S. (eds) (2013), *Tourism and Crisis* (London: Routledge).

Wall Reinius, S. and Fredman, P. (2007), 'Protected Areas as Attractions', *Annals of Tourism Research* 34:4, 839–854.

Westin, L. (2006), 'Trading Natural Resources for Public Grants: Development Rhetoric, Image, and Social Capital in North Sweden', in Kasuhisa et al. (eds).

Zachrisson, A., Sandell, K., Fredman, P. and Eckerberg, K. (2006), 'Tourism and Protected Areas: Motives, Actors and Processes', *The International Journal of Biodiversity Science and Management* 2:4, 350–358.

Zillinger, M. (2007), 'Organizing Tourism Development in Peripheral Areas: The Case of the Subarctic Project in Northern Sweden', in Müller and Jansson (eds).

Chapter 14

Responsible Tourism Governance.
A Case Study of Svalbard and Nunavut

Arvid Viken, Margaret Johnston, Torill Nyseth and Jackie Dawson

This chapter explores processes of tourism development governance in two Arctic regions: East Svalbard, Norway and Nunavut, Canada. Both regions are tourism destinations for relatively small numbers of tourists, but are nonetheless important for those involved in the businesses serving them and for their respective government authorities. Due to purely institutionalized tourism systems (Nunko et al. 2012), it can be discussed whether the areas in question really are 'tourism regions' or destinations (see Saarinen in Chapter 3 of this volume). But both areas are destination regions in the sense that they are affected by matters of politics, policies and governance. And in both regions tourism is expected to increase, and to create new governance needs. Developing destinations or tourist regions, as we prefer to call the two case areas, is a matter for various stakeholders (Baggio et al. 2010). Among the stakeholders are government authorities; in Arctic regions government stakeholders are probably more involved than in other regions, because of remoteness from markets, complexity of management and the need to preserve and protect vulnerable nature. Therefore, rather strict environmental regimes, primarily conducted by central or national authorities, are widely accepted in the eight countries that comprise the Arctic nations.

Yet Arctic residents are also part of a larger change in stakeholder governance. With new citizenry involvement in the western democracies, the legitimacy of strict regimes is being contested (Wallington et al. 2008). Local control over local resources has become an important aspect of governance, reflecting the belief that local communities have the rights and the knowledge required to manage their environments through democratic governing (Lane and Corbett 2005). Further, it is argued, many of the environmental problems have local origins and can best be solved with local involvement (cf. Gibbs and Jonas 2000). However, this also means that there is a plurality of interests and actors concerning development and environment issues, and a need for a stakeholder-based management approach. Multi-party involvement is said to be an important aspect of modern governance (Baggio et al. 2010). According to Beritelli et al. (2007, 96), '[t]he concept of governance applied to tourist destinations consists of setting and developing rules and mechanisms for a policy, as well as business strategies, by involving institutions and individuals'.

The Arctic needs responsible governance regimes that are able to accommodate both environmental needs and the development of industrial activities such as tourism. The question we are exploring is how models of governance influence the development of tourism region. In this chapter the question will be examined through a study of two different approaches to governing tourism development in two remote Arctic regions: Svalbard, Norway and Nunavut, Canada. The case regions are quite different, as they are in located in different countries and are in different stages of tourism development. However, in both regions tourism development issues are on the agenda, and subject to current matters of governance and management. Both regions cover large areas and have sparse populations. Nunavut has population of 34,000, comprising aboriginal and non-aboriginal individuals spread across 28 communities (Government of Nunavut 2013). Svalbard has about 2800 inhabitants separated in two major settlements, none of whom are aboriginal to Svalbard, and about 2000 of them live in the Norwegian community, Longyearbyen, situated 960 km north of mainland Norway. Svalbard has traditionally been governed through a centralized and hierarchical governance model, particularly in relation to environmental questions and, since 2002, in relation to local governing in Longyearbyen. East Svalbard, the specific area studied here, is not inhabited but is visited by hundreds of cruise tourists every year. Tourism development in Nunavut, on the other hand, has been governed through a less clearly defined approach, including both an emphasis on community involvement and an expectation of broader development leadership by government and the industry. Analysis in both cases is based largely on interviews with key stakeholders and participants in the planning processes.

Theoretical Point of Departure: Responsible Governance

Modern governance has emphasized the role of the state in managing the environment, natural resources and economic development through legislation and planning direction. However, the role of government has changed from 'command-and-control' (Pellizzoni 2005) to a role of enabling and cooperating (Kapoor 2001). The transition from governing to governance entails a shift in the dominant approach to responsibility (Pellizzoni 2004, 542). According to Wallington et al. (2008, 280), '[t]he shift away from state-based regulation and towards 'horizontal' governance arrangements intuitively involves a change in social relations of responsibility between the state and the civil society'. Responsibility is one of today's political master frames, but it is far from clear what it means exactly. It is a term used when people in authority make the 'right' decisions or take the 'right' actions in a moral sense (ibid.). Pellizzoni (2004) argues for responsibility as an applicable concept in the analysis of environmental governance, and presents a typology for such analyses. This approach has been followed up by several authors. Among these are Wallington et al. (2008), who consider two forms of responsibility: accountability and responsiveness. As they see it, "[b]eing responsible for one's conduct … .

not only implies that actions are voluntary and performed with full knowledge of the situation (Young 2006) but also that the consequences of the events can be anticipated and calculated" (Wallington et al. 2008, 280). 'Responsible' is a term originating from the Latin verb *respondere*, or to answer, or account for. Within this thinking, the individual is seen as a rational and autonomous agent, morally responsible for his or her actions (Pellizzoni 2004). This way of looking at responsibility requires a solid and defendable foundation for decisions such as scientific evidence and documentation of actions. Dimensions of responsibility are liability, accountability and responsiveness. Accountability, however, is a passive and reactive form of responsibility, according to Wallington et al. (2008), who, following Pellizzoni (2004), also include an anticipatory logic in responsibility, a 'reflexive, forward-looking complement to accountability' (ibid., 281). To Pellezzoni (2004) this is responsive governance, not only based on science and perceived truth, but also based on the inclusion of arguments, reasoning, values and meanings. And he continues:

> Response entails previous listening to a question. It entails openness, a willingness to understand and confront the other's commitments and concerns with ours, to look for a possible terrain of sharing. It entails readiness to rethink our own problem definition, goals, strategies and identity. We can talk also in this case of adaptation, … (ibid., 557).

Thus, responsiveness is to take the views and values of other actors seriously, 'having respect for the integrity of practices and the autonomy of groups; responsiveness to 'the complex texture of social life' (Seltznick 1992, 465)', as Braithwaite (2006, 885) notes. Thus, responsive governance refers to a system where those governed or managed take part in the creation and implementation of the management or governance regime, and where the regime is responsive towards established practices and traditions. Therefore, governance and management systems should be transparent and negotiable. In applying this model, no rational decision making model is at work; the bases for decisions are not causes, but reasons, as Pellizzoni sees it (ibid., 547). Whereas an accountability approach will focus on facts, figures and validity, responsiveness is based on platforms that are socially robust, reflecting the values and desires of the stakeholder citizenry. Given the nature of governing practices in many places today, accountability has to be complemented by responsiveness if an effective management regime is to be maintained. As claimed by Wallington et al. (2008, 285):

> The necessary complement to public accountability for the achievement of effective governance lies in a responsive public discourse that enables the contestation of problem framings, and takes seriously the socio-cognitive competence of lay publics and their capacity to reframe public issues. This lay competence is based on a thorough assessment of the quality of the

existing knowledge for policy, as well as on the quality of institutionalized social relationships.

Responsibility is also linked to legitimacy and trust, as legitimacy turns power into authority (Pellizzoni 2004). In the new landscape of governing no single agency is able to realize legitimate and effective governing by itself (Kooiman 2003). However, the legitimacy issue is dynamic and always subject to tensions, and without it a governance regime will be weak. According to Parkinson (2003), there are three dimensions of legitimacy: legality (legislation and regulation), justifiability (accepted as shared beliefs) and consent (support of the political system). All three aspects are part of and influenced by the processes studied here.

The legitimacy focus in relation to responsive governance raises the question of who is, or who should be, in charge of overall governance. Who should have the overall legitimate power? Responsible government can take place in a hierarchical, government based steering system, through emphasizing rational causes for the management suggested, or through a self-organized network-based and responsive governance system that is societally resilient, reflecting accepted values and goals. The state is the only real authority, being the ' ... source of democratic, accountable governance' (Pierre and Peters 2000, 197). However, when this status is extended to civil society, it is either proclaimed openly, or it is disguised in the parlance of stakeholder participation. Responsible hierarchical governance could be related to constructing the 'rules of the game', for instance, through institutional design, the medium through which actors can act and try to use these rules in accordance with their own objectives and interests. Management plan development can also be understood in this way. State actors often also play a role in governance through their control of legal, institutional and financial resources (Kooiman 1993). In systems of self-governance and shared responsibility, governance takes place through the establishment of a decision making and management model that reflects the agreement among stakeholders about what is an appropriate approach for the situation and what will be effective given the abilities and interests of the group members. How much real control can be transferred to the stakeholder group may be limited by the legal responsibility embedded in the existing hierarchical governing system.

In the following sections we describe the two cases in the light of this understanding of responsible governance. Do we find any forms of responsible governance in Svalbard and Nunavut? How are these forms of governance legitimized? How are multiplicities of interests and tensions dealt with?

East Svalbard: Contested Issues and Inclusive Planning Processes

The management regimes

Svalbard nature is viewed as both fragile and harsh. It has always been part of Norwegian politics to protect this Arctic environment, since Norway accepted management control of the archipelago through the international Svalbard Treaty signed in Paris in 1920. The exercise of management control requires demilitarization, equal economic access for the signatory parties, and protection of the environment. Reindeer and polar bear have been protected since 1925 and 1973, respectively. For cultural heritage, there is a general protection of artifacts, sites and buildings from 1949 or earlier. Heritage sites usually coincide with those locations where it is easy to go ashore, where there is a creek or river, and where there is vegetation. These are the places where whalers and trappers preferred to go ashore, and therefore have traces of former human activities. Thus, most of the places where tourism operators go ashore with their tourists are heritage sites and these sites are normally subject to some sort of protection regime.

Around 65% of terrestrial Svalbard is protected. There are several forms: national parks, preserved, but accessible for the public; nature reserves, preserved for the sake of nature, but normally accessible; bird sanctuaries; and, walrus breeding sites, where people have to keep 300 or 500 metres distant. Then there are some general protections, such as a prohibition on removal of plants (and picking flowers) and heritage. A proposal from 2005/2009 was implemented concerning heritage regulation: nine heritage sites were closed to access. For these sites, people are not allowed to go closer than 300 metres. The protection regime of East Svalbard has different justifications. There are several bird reserves and protected heritage sites that derived from the 2005/2009 closure proposal (valid from 2010). But the most important here are two nature reserves that are regulated through guidelines (*forskrift*): Northeast Svalbard and Southeast Svalbard. The aim for this protection framework is to preserve huge, unfragmented and untouched terrestrial and marine areas with intact nature, ecosystems, landscapes and heritage as reference areas for research (For 1973–0601 nr 3780). The areas, together with adjacent national parks, cover about 30% of Svalbard. However, the areas have been open for the public, and are visited by smaller and expedition style cruise ships (with an upper limit of 200 passengers set by Governor's Office rules).

To manage such areas there is a need for evidence of the environmental quality and character and for a system to monitor impacts and change. In these areas, this evidence is lacking completely, or what is available does not cover all required aspects. The Norwegian Polar Institute and Directorate for Nature, two parties concerned about the environment, science and protection, therefore, demanded a closure of the sites due to the lack of systematic monitoring in order to determine whether change was occurring. If activities were to be permitted to continue, it would be necessary to provide such a system. Those arguing against the closures asked for evidence of any problems posed by having activities in the area, claiming that there was nothing to indicate any negative issues. Others argued that some evidence related to impacts or change did, indeed, exist. For decades the Governor has made site inspections around the islands. It was argued that these monitoring assessments had been sufficient for viable management to date.

Regulations suggested in 2005 and a management plan suggested in 2013 both only concern the actual uses of the areas, not the broader protection goals and decisions. Among the issues in the planning process was the need to make operational the term 'reference area for research'. Some in the research community interpreted this term to mean areas where people should be excluded except for certain research purposes. The management plan does not follow this interpretation. In the management plan, the areas are split into three zones: a) areas closed off as reference areas: b) areas closed between May 15 and August 15 [1]: and, c) areas open for public and tourism activities in principle, but under some sort of self-control such as site guidelines and reporting.

The management plan process

There might be several reasons why the issue of appropriate management of East Svalbard arose in the early 2000s. The Norwegian Ministry of Environment created a working group at a 'directorate' level to judge 'the vulnerability of the different nature- and cultural values on Svalbard in relation to the growing cruise tourism, and ... to make a proposal of how the ship and cruise tourism should be regulated' (Internal note 2005, 1). The proposal was to forbid travel in most of East Svalbard, based on the argument that it should be strictly maintained as a reference area for research, that is, an area closed for all human activities except for research. The proposal was accepted by the Norwegian Inter-Ministerial Polar committee, a permanent ministerial committee that meets annually or semi-annually, and then passed on to the Governor of Svalbard for implementation. Although large parts of the area were already preserved as national reserves, the suggestion represented a major change in the management regime. Previously, the area was protected, but people were allowed access; this new management plan indicated that it should now be closed in practice, except for a particular number of visitor sites. The Governor of Svalbard sent the proposal to stakeholders and interested parties for their input.

The response to this proposal was massive, particularly from parties in the tourist industry. Due to ice conditions, the eastern areas normally were only accessible around the end of August and a few successive weeks. However, the loss of ice cover in the last decade has opened the area and made it into a much more interesting destination for the cruise industry. Therefore, the cruise tourism operators' organization, Association of Arctic Expedition Cruise Operators (AECO), protested loudly, and began to seek allies among many actors, from local residents on Svalbard to politicians in the Parliament and individuals from foreign embassies. The proposed regulation was interpreted as an exclusion of some research activities such as those of the local university studies (UNIS). It was also disputed publicly and was critiqued, primarily as a consequence of the way the issue was launched, as not having any democratic process as its basis. In particular, the local newspaper published opposing views, mediated by the editor and several readers' letters. The Governor of Svalbard was not pleased with the

---

1   These are one smaller island (Langåra), and a rather large archipelago called Tusenøyane (*The Thousand Islands*).

negative attention and response, and after several years of subsequent hearings, it was decided to put the proposal aside and to start an alternative and more inclusive process to develop a management plan for East Svalbard. Management plans are a well-known conservation tool in Norway, and a standard tool for stakeholder involvement, clarifying how a preserved area can be used and how the conservation decisions are to be interpreted. The Governor, of course, had observed the protests, and it was decided to include the major interests in the process. Four working groups, covering the most important interests and stakeholders, were set up; one from the science community, a tourism industry group, a fishery group, and a local community group. As part of their decision platforms, according to a listing on the home page of Governor, 14 expert reports were produced. In addition, a reference group accountable to the Governor discussed the proposals from the working groups and it made recommendations of components that should go into the management plan. The Governor put together the suggestions from the groups and made an encompassing proposal. As part of a normal procedure, a hearing was undertaken, during the summer and autumn 2012, and 32 hearing responses were submitted. After this, the draft management plan was sentto the Directorate for Nature Management, eventually to be handed over to the Ministry of Environment for final endorsement.

Responses through the hearing process (accessible on the Internet) and focus group interviews with representatives of the stakeholders reveal a number of points of interest. First, it is not clear what the motives for the suggested closure and the subsequent management plan are, and there is some confusion concerning the environmental status of the area. Second, the case shows that there are different views about how environment-tourism relations should be managed. Third, there are concerns about what particular terms mean and how they are being used. The term 'reference area' was widely held to be vague, as was the term 'to be precautionary' (*føre var*); it is also generally unclear what is meant by a 'knowledge base' for viable management. Fourth, there are very diverse opinions of the performance of the planning process and whether it has been democratic or rather manipulative. Thus, planning as a governance process seems to have been a struggle between traditional hierarchical steering and accountability and the responsive approaches of network based governance and meta-governance.

Stakeholder participation

Concerning the work in the stakeholder groups, the local people are divided in their views. The process seems to have been smooth and democratic, and it involved all relevant interests, according to the respondents we approached. The result is also something most people seem to be able to live with. One of those representing the local interests stated that it was wrong to make rules hindering local people from visiting areas that until now had been accessible; rather, an exception should have been made for local residents. However, he also admitted, that there were strong symbolic elements in this view, a reduction of the encompassing

'Svalbardian' freedom. Symbolically, a state-imposed regulation also tends to be seen as counteracting the newly obtained local democracy (effective from 2002) and, as one interviewee claimed, equal to dictatorship in the local public eyes. Two representatives of the tourism industry did not feel that this had been a good process. As a member of the tourism stakeholder group, one reported feeling like a 'hostage', believing that those in charge were not really willing to listen to his views. As he saw it, those in charge had already made their decisions and used the working groups as a 'democratic alibi'. A representative of the largest local tourism company argued similarly. He felt that he and the local people had been 'cheated'. Regarding the actual result, the management plan itself, the tourism representatives were hesitant about the two closures. But, overall, the plan is seen as a victory for those who protested against the closure first suggested in 2005. The major difference between this process and the first amendment, therefore, relates to the process: the management plan has been constructed, at least on the surface, through an inclusive and democratic process.

But the management plan also takes into account the work the tourism industry has undertaken to stand out as a responsible actor. The self-regulation system that is built into the plan – site guidelines, announcing of itineraries, and reporting to the Governor afterwards – is an adoption of an existing regime. On its own initiative the tourism industry has developed site-specific guidelines for many sites, developed in collaboration with natural science and heritage experts (Association of Arctic Expedition Cruise Operators n.d.). In this sense the industry has performed an *ex ante* responsibility. This is also acknowledged by the Governor and by the central authorities, where a representative says that:

> The tourist industry at Svalbard is a responsible industry. We go for a close collaboration with the industry. We believe that through collaboration we will increase the industry's understanding of what protection is for, which also affect the product, the tourist adventure. Our experience with this form of collaboration is positive. People like responsibility, to be trusted.

Through these forms of collaboration among government, environmental experts and the industry a common ground of understanding has been established.

Nunavut: Finding the Fit for Tourism in Economic Development

The management regimes

The territory of Nunavut, Canada was established in 1999 through a division of the Northwest Territories, following a land claims agreement with the Government of Canada finalized in 1993. The boundaries of Nunavut encompass the traditional lands of the Inuit people and the territory's creation reflects the desire of Inuit for self-governance of their communities and resources. Separation from the

Northwest Territories meant that Nunavut had to create its own political system, legislation, regulations, and management structures. The territory's first economic development strategy was produced in 2003. The strategy stated that the 'Nunavut economy is far behind other jurisdictions in Canada' and noted that almost all economic fundamentals needed to be established before a 'thriving, diverse, business – and community – driven economy [could] be created' (SEDS 2003). The report highlighted an extremely limited physical infrastructure, an 'under-skilled' workforce, under-developed essential services, and regional policy and regulatory frameworks in early stages of establishment and implementation (SEDS 2003). The strategy highlighted three sectors seen to offer excellent prospects for large-scale economic growth: minerals, fishing, and tourism.

As the territory began to develop, the priorities of the Government of Nunavut were not with tourism, but rather with other means of economic development, especially mining and fisheries, and with services such as education and health (Robbins 2007; Government of Nunavut 2010). The land claims agreement that established Nunavut does not address tourism development specifically, but it does address mining, oil and gas, resource development; this is similar to the latest territorial economic outlook report, which highlights mining and fisheries (Nunavut Economic Forum 2010). Progress has been slow for tourism, but for the other sectors the settlement of the land claims meant that industrial development could begin, especially mining (Johnston et al. 2012). The mining sector has seen major exploration, development and regulatory support, reflecting government commitment to providing a strong management regime (The Northern Miner 2009).

Despite early planning efforts, the same issues have plagued tourism development for several decades: building community capacity and tourism awareness, product development, access to markets and finance, and training (Robbins 2007; Johnston 2011). Further, the governance framework for tourism management has been unclear to stakeholders. A number of tourism decision makers and regulators interviewed in 2010 indicated that the responsibility for tourism governance had been unclear for some time (Johnston et al. 2012). Prior to the creation of Nunavut, the Northwest Territories government had managed licensing and regulation, but in the mid-1990s created Nunavut Tourism, an industry membership organization that was given responsibility for marketing, product development and training (Robbins 2007). The planning, licensing and enforcement aspects of governance were retained by the government. With the creation of Nunavut in 1999, it appears that these responsibilities were not prioritized by the government, leading to confusion about the governance regime and, eventually, capacity issues in terms of management actions (Johnston 2011; Johnston et al. 2012).

Stewart, Draper and Dawson (2011) note that given a lack of high-level support for tourism development in Nunavut, tourism growth has been slow and inconsistent, with an *ad hoc* character, only overcome by community enthusiasm and support from sub-sectors of the industry such as the cruise ship industry. The majority of tourists are business travellers (Nunavut Tourism 2011), but leisure travel is predicted to grow over the next decade, particularly expedition cruise and pleasure craft activity, largely because of changing ice conditions (Stewart et al. 2010; Stewart et al. 2013). The absence of a strong management regime, combined with environmental challenges related to climate change, has meant that tourism in Nunavut is in a constant state of flux (Stewart et al. 2011). Further, until recently, the Government of Nunavut policy of decentralization meant that the tourism unit was located in Pangnirtung, rather than in the capital where many of the other territorial and federal government departments and Nunavut Tourism are located.

The management plan process

Early tourism planning in the geographic area that became Nunavut Territory followed what is known as the community based approach, a locally-focused approach that appears well suited for the small, air and water accessible communities of the territory. This approach is predicated strongly on local involvement that hears all views in the development of a plan based largely on values and traditions. Pangnirtung was the first community to participate in the process and was followed by other communities in the region through the 1980s (Robbins 2007). An attempt at strategic tourism planning for the territory as a whole began with the Blackstone Report, a 2001 document titled 'A Strategic Plan for Tourism Development in Nunavut', a report which also noted the heavy prevalence of business travellers (Robbins 2007). Tourism at this time was seen as 'a good fit for Nunavut culture and communities' (SEDS 2003) and was viewed as supporting Inuit culture and providing opportunities for more stable community-based economic activity than the dramatic boom and bust cycle experienced in natural resource extraction industries (SEDS 2003) . Indeed, after the closure of Nunavut's major mine project in 2002 (Nanisivik), tourism was thought to be the territory's 'single most important economic activity in the private sector, in terms of its contribution to the Territory's GDP' (SEDS 2003). This was at a time when the vast economic potential of mineral activity was not yet clear.

No tourism sector strategy came out of this early work, but another attempt was made in 2007, repeating the community consultations that were the foundation in 2001, reflecting an ongoing commitment to responsive governance and a readiness to listen to all views. Sector development proceeded on an *ad hoc* basis without a strategy, relying on the marketing and training activities of Nunavut Tourism, the product development of the industry itself, and community willingness to participate. Despite the publication of a document outlining tourism resources and sector potential, no strategy was developed after these efforts. In 2011 work began again to develop a tourism sector strategy, this time with some success. Interviews were undertaken with the main players in the most recent process and the material below arises from those sessions.

The 2012 tourism strategy: Moving forward

As the other sectors moved forward and grew, supported by economic development strategies and an appropriate governance framework, tourism suffered and did not appear to be meeting its projected potential. Despite the earlier attempts to develop a tourism strategy and extensive consultations, an actual strategic plan had not been created. Consultation with residents of communities has always been part of the tourism planning efforts, but the processes in 2001 and 2007 were halted prior to the establishment of the strategy. In 2011, consultations also occurred as a precursor to the development of a strategy by a broad group of stakeholders led by the Government of Nunavut's Department of Economic Development and Transportation, the body responsible for regulation and planning of tourism. It was expected that a tourism strategy would be launched at the March '2012 Tourism in Nunavut Conference' as a symbol of the sector's renewal, but the planning process stalled and deadlines were not met to enable this launch.

After three false starts and extensive and repeated efforts to consult with local residents, tourism operators, policy makers and other stakeholders, no tourism strategy had been established. On each occasion, according to one interviewee, strategy development was halted by individuals who seemed to 'single-handedly de-rail the process', ultimately because of existing 'politics' or conflict between stakeholder groups or, in some cases, individuals within stakeholder groups related to differing opinions on the content and approach of the draft. When it became clear that the 2011 process was repeating historic patterns of political stagnation, a senior official in the Government of Nunavut forced a new approach in an effort to finally produce a tourism strategy for government deliberation and decision making.

The new process resulted in a draft document being completed in early 2013, presented for government departmental review and eventual consideration by the cabinet and the government legislature in spring of 2013 for approval and implementation. The long-awaited strategy focuses strongly on the notion of partnerships and recognizes that 'the development of a successful tourism industry will be a long and complex process' that requires coordination and commitment in all sector activities. In addition to strong partnerships, the tabled strategy emphasizes the need for strong legislation and regulation, quality attractions, produces and services and superior education and training. The new strategy includes goals, action plans, and a proposed budget that includes potential sources of funding. Stakeholders involved in the development of the strategy had mixed opinions on how realistic the action items are considering the precarious and constantly changing nature of the federal and territorial funding environments.

Stakeholder participation

The revised process implemented by the senior government official to generate the new strategy involved a more focussed group of stakeholders than was brought together in 2011. Six core groups were the most involved in tourism development and management in Nunavut: Tourism and Cultural Industries (a department of Economic Development and Transportation (EDT) in the Government of Nunavut), Canadian Northern Development Agency, Nunavut Tungavik Inc. (NTI), Arctic College, Nunavut Tourism, and Parks Canada. The individuals chosen to represent these institutions were long-term residents of the territory and had experienced the failure of previous attempts to develop a strategy. A seventh group, Nunavut Community Economic Development Organization, was included in later discussions. On the insistence of the senior government official, an external facilitator was hired to lead, facilitate and, for the first time, bring the process to successful completion. This group worked together with the facilitator to create a document and obtain consensus, and then present the document more broadly in the community for input.

Though there was consensus on the content of the document, there was not unanimity, according to our interviewees. Most stakeholders were of the same mind in terms of the approach to key themes and, to some extent, financing. It appears that the current version covers the appropriate areas, but, as mentioned, some concerns have been raised about costs. One interviewee said the current version contained no more 'magical money', meaning that the financing approach was now more realistic compared to the 2011 draft, yet another said that the current version of the plan required too much money with no total amount, and so it would likely not be approved by the government. To avoid unwanted delay due to slow bureaucratic federal processes, the facilitator, in consultation with the senior government official, decided that only three of the seven stakeholder groups would act a signatories on the final document (EDT, NT, and NTI).

The creation of a strategy is seen by the interviewees as important in establishing the role of the tourism sector in economic development in the territory. This is the final economic sector to complete a strategic plan and, in some cases, other sectors are working on their third five-year strategies. Concerns were noted that the tourism sector itself has not been viewed as contributing a great deal to the economy, in part because there has been no strategy, but also because there were problems related to the operations of Nunavut Tourism prior to 2008 that had to be overcome and the fact that the territorial tourism department was not in the capital. In addition, senior decision makers in the Government of Nunavut have focused heavily on the development of mining activities believing that the resource sector is more promising in terms of economic prosperity and self-sufficiency than tourism.

Discussion

The perspective given as a point of departure in this chapter is responsible governance related to tourism development in two Arctic regions. Originally a non-restrictive approach was taken in tourism development in both regions, but the two regions represent different phases of development of the tourist industry. In Svalbard, tourism has been a strategic priority area since 1990, and the industry has been growing ever since. With growing cruise tourism in the eastern regions of the islands, a strict regime was suggested – the area should be closed for tourists. When the authorities in Svalbard withdrew this suggestion (2009), substituting it with a stakeholder management plan process, this might be understood as a transformation from governing as accountability towards governing as responsiveness. The opposite happened in Nunavut. When the senior government official intervened in 2012 to push the development of a tourism strategy in Nunavut forward, the process became less one of collaboration and participation and more one of heightened accountability based on a clear hierarchy. However, in both cases studied, governance involving a collaborative, shared responsibility model has been a major part of the process.

In both cases, the planning process has been based on involvement and stakeholder participation, a process not focused merely on scientific evidence, but also on local acceptance and a more general openness similar to what Pellizzoni has observed as 'a receptive attitude to external inputs to help in deciding what to do' (2004, 557). In such processes the arguments and stakeholders are reciprocally confronted and judged, 'harnessing the human capacity of reflexive agency' (Wellington et al. 2008, 286), and strengthening ideas of shared values and meanings. To obtain this, meaningful dialogue is a requirement.

Another important aspect is to take into account the management systems that already exist. If there is an existing responsible regime, there might not be need for a new one. As Braithwaite (2006, 886) claims, 'law enforcers should be responsive to how effectively the citizens or corporations are regulating themselves before deciding whether to escalate intervention'. Self-governance refers to the capacity of social entities to govern themselves autonomously (Kooiman 2003,79). This is not new and has existed as a part of governance systems for centuries, for instance the self-governance of local councils. Today, the types of selves targeted by governance interventions are not only other public authorities, but also communities, interest organizations, businesses and policy networks whose self-governing powers are invoked in order to make them more effective, efficient and democratic. This principle is followed in the management plan for East Svalbard: the site guidelines produced by AECO are a part, as are the cruise industry's organization and self-reporting systems. This strengthens the perception of a democratic process and is a way of gaining legitimacy. In Nunavut, the entire tourism industry has been self-organizing for over a decade and has done so with some success. Even without a tourism strategy and in the absence of self-reporting and self-regulation systems and institutions such as AECO, the territory was, through efforts of Nunavut Tourism, the Government of Nunavut (EDT) and other stakeholder groups, able to increase reported revenues by 40% between 2008 and 2011.

Stakeholders involved in governing or managing tourism in Svalbard and Nunavut made use of terms like *acting responsibly* and *adaptive management.* However, not all thought of the processes as responsiveness. Particularly in Svalbard some industry stakeholders perceived the participation and the openness as simply the old model being given a new wrapping. The authorities appeared split in their views of the tourism industry. Shared responsibility depends on trust between the governor and the governed. Some authorities have no trust in the industry's good will and ability to self-organize, while others look upon the sector as responsible and accountable. The Svalbard case described above stands out as a battle between two modes of governance, the traditional hierarchical model which still is the basic model, here called accountability, and new forms of more reflexive and responsive governance regimes. The old model did not have a legitimate position any more. In Nunavut it was slightly different. Here the process was in some ways so responsive and collaborative that it essentially caused a decade of stagnation. For example, when consensus was repeatedly not reached among the large range of stakeholders, for a variety of interpersonal and political reasons, no single point of authority was present to 'force' the development of an admittedly imperfect strategy even for the sake of progress. But also this can be seen as lack of legitimacy, no one collaborator was in a position to enforce a strategy for development.

In relation to Parkinson's (2003) approach to legitimacy, there are several differences between the two cases, and some similarities. Concerning legality, legislation and regulation, in the Svalbard case, the existing rules were accepted and a foundation for the management planning process. However, there appeared to be some doubt about the Governor's ability as a regulator: as viewed by both Norwegian Polar Institute and the Directorate for Nature, the Governor is suspected as having too close relations with the local community. This is similar to what is found other places; the regional level is not furnished by traditional platforms of authority (cf. Wallington et al. 2008, 13). In Nunavut, a history of interpersonal and political conflicts has clearly affected the region's ability to maintain its legitimacy through legality. Additionally, the limited tourism specific regulation and legislation and, until 2013, no tourism strategy that would support tourism growth in the territory have made encouraging economic development based on the industry a challenge.

In discussing justifiability as a form of legitimacy, Wallington et al. (2008), operate with a split between input-oriented and output-oriented legitimacy. In the Svalbard case, as with the Nunavut case, input-oriented legitimacy is obtained through the stakeholder approach; nobody was criticizing the involvement (input) model chosen. However, concerning the output, when some stakeholders in Svalbard perceived their participation as being manipulated rather than honestly sought, it caused a severe stress in the new form of governance. In Nunavut, outputs were also cause for significant turmoil between stakeholder groups and, ultimately, reflected in decade long stagnation of tourism development planning and management.

Consent related legitimacy is the third Parkinson's category. This is limited in both cases. In fact, some actors in Svalbard probably never had confidence in the system, or in the motives of the authorities. There is a recent tradition for a network based governance (Viken 2011), but there is an even stronger tradition for old-fashioned hierarchical governing. In Nunavut, because the system was much more flexible and participatory from the beginning there was high consent among stakeholders in the process and particularly in the desired goal to develop a tourism strategy. However, like in Svalbard, there remained trust issues and power struggles between stakeholders groups and government departments, including a feeling of being dismissed by the federal and territorial governments given the allure of natural resource extraction opportunities.

Conclusion

The contrasting and corroborating case studies described in this chapter share a tale of the challenges associated with tourism governance in remote and fragile regions of the globe. The cases demonstrate that progress towards a destination status involves politics, planning and governance, and the more encompassing analytical perspective of a tourism region, that Saarinen (2004; Chapter 3 of this volume) suggests, is shown to be viable. Despite the distinctions in the approaches taken in both Svalbard and Nunavut, the importance of responsible governance is clear. In a sector that relies so heavily on partnerships, collaboration and cooperation (in the areas of transportation, attractions, services, accommodation, etc.), the importance of a participatory and flexible approach is self-evident. Exactly how the process transforms reality, however, can be very different as we have seen in the cases of Svalbard and Nunavut. There is no 'one-size fits all' approach to governance and management. The regimes have to be contextual and situational. Participatory approaches can be more vital during some parts of a development process than in others. Additionally, Svalbard nature is often regarded as a global common good and should be governed as such. Nunavut, on the other hand operates primarily at the regional and local levels in respect of tourism and environment, with community involvement supporting Inuit culture as an important element. Thus, it could be asked whether responsiveness concerning East Svalbard – with no native population and no settlements in the area in question – should be as important as in Nunavut with an indigenous population. And further, is responsiveness based on stakeholder involvement the most really responsible governance? Could it be that tourism industry development should be a matter of responsive management, as it has been both in Svalbard (Viken 2011) and in Nunavut, whereas environmental preservation to a greater extent should be managed using the accountability model? The two models should not be seen as mutually exclusive, but rather as principles that have to be tuned towards each other, adapted to the tasks in question. In most cases, one could probably search for a compromise. Evidence, assessments and knowledge are important premises for governing decisions and plans. But without stakeholder 'buy-in' and appropriate consultations with local residents and industry, it will be challenging for government to implement management strategies as they will lack the social robustness that is required for effectiveness. As noted by Lemelin et al. (2012), the adaptive nature of the governance approach is also vital considering the constant change and the inevitable emergence of unforeseeable events such as are experienced regularly in the Arctic and in the tourism industry.

References

AECO (Association of Arctic Expeditions Cruise Operators) (2013), <http://www. spitsbergen-svalbard.com/2012/01/04/aeco-site-specific-guidelines.html>.

Baggio, R., Scott, N. and Cooper, C. (2007), 'Improving Tourism Destination Governance: A Complexity Science Approach', *Tourism Review* 65:4, 51–60.

Beritelli. P., Bieger, T. and Laesser C. (2007), 'Destination Governance: Using Corporate Governance Theories as a Foundation for Effective Destination Management', *Journal of Travel Research* 46:1, 96–107.

Braithwaite, J. (2006), 'Responsive Regulation and Developing Economies', *World Development* 34:5, 884–898.

Gibbs, D. and Jonas, A.E.G. (2000), 'Governance and Regulation in Local Environmental Policy: The Utility of a Regime Approach', *Geoforum* 31:3, 299–313.

Government of Nunavut (2013), *Nunavut Bureau of Statistics* <http://www.stats. gov.nu.ca/en/home.aspx> (home page), accessed 13 April 2013.

Johnston, A. (2011), *Stakeholder Perspectives on Change and Adaptation in Expedition Cruise tourism in Nunavut.* Unpublished master thesis. (Thunder Bay: School of Outdoor Recreation, Parks and Tourism, Lakehead University).

Johnston, A., Johnston, M.E., Stewart, E.J., Dawson, J. and Lemelin, R.H. (2012), 'Perspectives of Decision Makers and Regulators on Climate Change and Adaptation in Expedition Cruise Ship Tourism in Nunavut', *The Northern Review* 35, 69–85.

Kapoor, I. (2001), 'Towards Participatory Environmental Management?', *Journal of Environmental Management* 63:3, 269–279.

Kooiman, J. (2003), *Governing as Governance* (London: Sage).

Lane, M.B. and Corbett, T. (2005), 'The Tyranny of Localism: Indigenous Participation in Community-Based Environmental Management', *Journal of Environmental Policy and Planning* 7:2, 141–159.

Lemelin, R.H., Johnston, M., Dawson, D., Stewart, E. and Mattina, C. (2012), 'From Hunting and Fishing to Cultural Tourism and Ecotourism: Examining the Transitioning Tourism Industry in Nunavik', *The Polar Journal* 2:1, 39–60.

Luck, M. et al. (eds), *Cruise Tourism in Polar Regions: Promoting Environmental and Social Sustainability* (Earthscan: London).

Maher, P.T. et al. (eds) *Polar Tourism: Human, Environmental and Governance Dimensions* (New York: Cognizant).

Nunavut Economic Forum (2010), *Nunavut Economic Outlook 2010,* found at: <http://www.landclaimscoalition.ca/pdf/Nunavut_Economic_ Outlook_2010.pdf>.

Nunavut Tourism (2011), *Nunavut Visitor Exit Survey 2011* (Iqaluit: Nunavut Tourism and Government of Nunavut).

Nunkoo, R., Ramkissoon, H. and Gursoy, D. (2012), 'Public Trust in Tourism Institutions', *Annals of Tourism Research* 39:3, 1538–1564.

Parkinson, J. (2003), 'Legitimacy Problems in Deliberative Democracy', *Political Studies,* 51:1, 180–196.

Pellizzoni, L. (2004), 'Responsibility and Environmental Governance', *Environmental Politics*13:3, 541–565.

Pierre, J. and Peters, G. (2000), *Governance, the State and Public Policy* (London:Macmillan).

Robbins, M. (2007), 'Development of tourism in Arctic Canada', in Snyder and Stonehouse (eds.).

Saarinen, J. (2004), ''Destinations in Change': The Transformation Process of Tourist Destinations', *Tourist Studies* 4:2, 161–179.

Selznick, P. (1992), *The Moral Commonwealth: Social Theory and the Promise of Community* (Berkeley, CA: University of California Press).

Snyder, J.M. and Stonehouse, B. (eds) (2007), *Prospects for Polar Tourism* (Wallingford, CAB International).

Stewart, E.J., Draper, D. and Dawson, J. (2011), 'Arctic Human Dimension. Coping with Change and Vulnerability: A Case Study of Resident Attitudes Toward Tourism in Cambridge Bay and Pond Inlet, Nunavut, Canada', in Maher et al. (eds).

Stewart, E.J., Draper, D. and Dawson, J. (2010), 'Monitoring Patterns of Cruise Tourism across Arctic Canada', in Luck et al. (eds).

Stewart, E., Dawson, J., Howell, S., Johnston, M., Pearce, T. and Lemelin, R. (2013), 'Sea Ice Change and Cruise Tourism in Arctic Canada's Northwest Passage: Implications for Local Communities', *Polar Geography* 36:1/2, 142–162.

SEDS (Sivummut Economic Development Strategy) Group (2003), *Nunavut economic development strategy: Building a foundation for the future.* Retrieved from: <http://www.nunavuteconomicforum.ca/public/files/strategy/ NUNAVUTE.PDF>.

Viken, A. (2011), 'Tourism, Research and Governance on Svalbard: A Symbiotic Relationship', *Polar Record* 47, 335–347.

Wallington, T.J., Lawrence, G. and Loechel, B. (2008), 'Reflections on the Legitimacy of Regional Environmental Governance: Lessons from Australia's Experiment in the Natural Resource Management', *Journal of Environmental Policy and Planning,* 10:1, 1–30.

Young, I.M. (2006), 'Responsibility and Global Injustice: A Social Connection Model', *Social Philosophy and Policy* 23:1, 102–130.

Chapter 15

Epilogue: Reflections on Tourism Destination Development

Arvid Viken

Social scientists engage in studies of industrial development of many kinds, and the tourism industry is one of them. One of the particularities of social science language is its inherent capacity to see beyond an industry's economic meanings and to unveil the richness of actors, materials and practices as well as the values and rationalities that are tied to it and that enable or disable its development. Such approaches challenge conventional notions of an industry like tourism. This is why social scientists' way of looking at tourism may lead to a – for some annoyingly banal, but for many scholars demanding and fundamental – question like 'what is tourism?' The authors of this anthology have focused on the term 'destination', motivated by its widespread but often blurred and non-reflexive use among those who relate to tourism. The assumptions that mark academic as well as political, governmental and industrial usages of the term have been challenged through conceptual and empirical explorations from a variety of angles. This has produced reports on current conceptual and empirical trends – i.e. about turns and tactics – in the field of tourism.

An independent and crucial aim directing work with the book has been to contribute knowledge that may spur reflexivity among all those involved in tourist destination research and development. The aim has been to unveil biased and problematic aspects of destination development and bring the phenomenon out of the sphere of assumptions where it is naturalized and 'hidden' in seemingly objective texts (Tribe 2005). When considering the book as a whole, we approach this aim through a double strategy. First, three chapters are dedicated to a specific conceptual exploration of destinations. Tourism scholars are tied to the schemata of tourism studies, which direct their ways of defining and discussing tourism destinations. Similar to other fields, the field of tourism studies has its 'own cultural logic and attended with associated forms of "capital" or cultural resources' (Fries 2009, 330). Such fields are conditioned by the scientist's background i.e. by their 'capital' and 'habitus', to phrase it in Pierre Bourdieu's terms. Capital is not only economic, but also cultural, social, physical and symbolic (Bourdieu 1986), facets which are in turn embodied in habitus. Habitus is inscribed in 'cognitive schemata that orient behavior in terms of beliefs about the nature of social reality' (Fries 2009, 330). The idea behind the conceptual part of the book has been to apply the reflexive practice of revealing parts of the schemata tied to different ways of defining and discussing tourism destinations.

Reflexivity can be defined in this and many other ways (see for example Bourdieu and Wacquant 1992; Alvesson and Sköldberg 2000; Lynch 2000; Alvesson et al. 2008). Alvesson and colleagues see it as 'the institutional, social and political processes whereby research is conducted and knowledge is produced', but it can also be 'to explore the situated nature of knowledge' (2008, 480). While working within mainly qualitative research traditions and from different social constructivist positions, the diverse authors of this book relate differently to the many facets of reflexivity. Thus, a second way of turning towards the reflexive aspect relates to the three empirical sections of the book. Here, destinations have been explored through identifications and explorations of a variety of processes, and expressed through the multiple voices involved (Alvesson et al. 2008). Such processes have been categorized as those of theming, reorientation and the politics of destinations. The chapters are written by scholars from different social science perspectives and hence bring in multiple perspectives that put the emphasis on a variety of processes involved in destination development. Sociologists, social anthropologists, geographers, planning scholars, and political scientists all apply theoretical approaches that are embedded not only in the field of tourism studies, but also in different branches of social science. In one way, the reflexive contribution of these chapters is their sum: as Alvesson et al. claim, '[it] is the accumulation of ... perspectives that amounts to reflexivity' (2008, 483). Here, reflexivity is present as a multi-perspective practice (ibid., 482–483), implying that the chapters overall provides a multifaceted understanding of the phenomenon of tourist destinations, especially those which can provide the basis for reflexive considerations of what destination development is and how it could be managed in appropriate ways. Thirdly, also an alternative way of looking at reflexivity is used, focussing on its analytical depth, as suggested by Kögler (1997), and fourthly the findings of the book are related to the partiality and situatedness of the knowledge that the book provides.

The Tourism Destination Discourse

As described above, the first section of the book is devoted to discussions of theoretical approaches to the field of tourist destinations. The section presents critical accounts concerning concepts, approaches and prevailing discourses of destination development, all adding to the reflexivity in the field by unfolding the multiplicity of the term, and the hybridity of the phenomenon in question. In Chapter 2, Arvid Viken presents some of the most common theoretical approaches found in the tourism literature. A discourse analytical methodology is applied, uncovering master narratives behind, and hegemonic positions within tourism destination theories. Viken demonstrates that in most theoretical stances the growth paradigm is taken for granted. Thus, business oriented discourses and theories hold a hegemonic position within this academic field, and partly inflict social science oriented analyses. These theories, terms and perspectives are taught in universities and colleges, as well as in vocational training, and hence dominate the parlance or rhetoric of tourism within the industrial field. One of the consequences is that the situation and challenge for SMEs and family and life-style based performers in the industry are scarcely covered. In the next Chapter 3, Jarkko Saarinen also sees destinations as socially constructed and as emerging from particular discourses. Saarinen ties destination to uses of the term 'region' (see also Saarinen 2004). A region is 'strongly related to the social production of space', he argues, and is influenced by political and administrative systems and processes as well as formed by globalization and homogenization trends. Saarinen argues that homogenization processes often perform locally within enclaves, both including and excluding people, and therefore represent a differentiation of place, producing and preserving a split between wealthy and poor people. Thus, tourism destinations are part of international or global discourses that tend to empower those in the centre while disempowering actors in the peripheries, which are in turn often destinations for those in the centre. This pattern is socially and culturally contingent and difficult to emancipate; in a global perspective, this is a matter of path dependency, to use Saarinen's own term. However, hegemonies are constantly challenged and niches always exist. It could be added that tourism has some sort of anarchistic character that provides space for those that do not perform inside or adapt to comprehensive international tourism systems. This is very much the message of Simone Abram who, in Chapter 4, describes tourism destinations as social systems to which the individual is socialized through family life, education, work, and leisure. Tourism is all over; it transforms us into tourists and into more or less uncritical spectators of processes through which places, regions and countries are turned into tourism destinations. In other words, there is a multitude of discourses and semiotic systems that adhere to tourism. As Abram sees it, it is an important part of tourism analyses to deconstruct such manifestations of power. Among the players within the tourism sphere are private businesses, large tour operating ventures, private organizations (for example NGOs) and a series of authority organizations. Abram also sees tourism development as a matter of planning, politics and infrastructure development. Such matters are often undertaken in ways that favour some while disadvantaging others. Most projects tend to be presented as an inevitable path to growth. However, Abram sees planning as a ritual process. Rituals tend to suspend traditional statuses, and in this way planning can make the status of a site uncertain and an occasion for challenging policy ambitions. To cope with tourism development, as a critical citizen, is something that can be and should be learned in order to immerse democracy into tourism development. In the final chapter of this section, Chapter 5, Brynhild Granås, looks at destinations from a perspective of space and place. Social constructivist and geographical theory development within tourism studies have rejected the term destination or reserved it for matters of political and managerial development. As a comprehensive and encompassing idea in tourism, Granås however argues for the relevance of holding on to the topic of destination. Her suggestion is to integrate Massey's place theory in tourism studies in a meticulous and premising way, making it a method of analysis. Relational processes where the place is performed as a destination for tourists – i.e. where places are destinized – are proposed as a focal topic of interest in destination research. Thus, the topic of destination is verbalized and subordinated to place. Her idea is to provide an integrative perspective that can attend to the empirical integratedness of for example the discrete and strategic of tourism development. Granås position the approach methodologically in an embodied and materially embedded social constructivism that emphasizes the relational and performative about production of space and place. Within this approach, we are enabled to observe how places are practiced as destinations – how they are destinized – and the complex and power relevant about tourism development seen from a social science perspective, is illuminated.

Themed Destinations

Theming is the headline for the first block of studies included in this book. As with other consumption practices, tourism seems to become more specialized over time and theming is a way of ordering these processes. But the series of possible themes are of course endless. Themes related to a place or region are catalysts of tourism. In this book there are explicit examples of how particular themes have given rise to a certain type of tourism development. The first case of theming, presented by Gunnarsdóttir and Jóhannesson in Chapter 6, is taken from history, about witch-burning in Strandir, Iceland. In Chapter 7, Arvid Viken refers to a sports activity, downhill skiing, a well-known theme of northern regions, particularly Finland. This has been copied in north Norway, but without significant success, as demonstrated in the chapter. In a third chapter on the topic, Chapter 8, Kari Jæger and Arvid Viken present an example of theming related to another sport, dog sledding, a theme that is interwoven with other themes as outdoor recreation and in the history of Arctic exploration. Thus, themes as platforms of destinations, normally related to networks of terms placing the destinations within a wider context.

There are of course different ways of categorizing platforms of tourism destination theming. One category discussed by Chang (2000, 36) is labelled 'place development themes'. Many places are pretty well known for particular attributes. Therefore, place names often connote themes. The North Cape, as other 'capes', is an example, as is the Arctic. This book does not include particular studies with Arctic as the theme. Still, the chapters about tourism development in Norrbotn in Sweden (Chapter 13), cruise tourism in the Lofoten Islands (Chapter 10) and about sled dog tourism in Finnmark (Chapter 8), all see Arctic nature as a theme that attracts tourists. What is explicitly focused on is sled dog tourism, discussed in Chapter 8, which is based on *polar* dogs. In this chapter, Jæger and Viken also raise the question of whether the Arctic wrapping really makes a difference. In Chapter 11, Eva Kajan and Jarkko Saarinen indicate that this is not enough, or that the destinations in their study do not put enough emphasis on renewal or path creation within this frame.

Chang (2000, 36) presents other ways of theming, such as through the marketing and branding of places. Marketing themes tend to be expressed through slogans, catchphrases and marketing material. In Målselv in northern Norway, studied both by Arvid Viken in Chapter 7 and Anniken Førde in Chapter 9, brands and slogans related to the Snowman have been applied and have infused the strategic thinking, planning, and product development of the resort. In several of the destinations studied, another destination theming category, 'theme park and attractions' (Chang 2000), are found – in Andøy and Målselv (north Norway) and in Strandir (Iceland). This is a sort of tourism development whereby a particular tourism is created around the structure of attractions (see Viken in Chapter 2 of this volume).

Local Reorientation

As stated at the beginning of this book, Arctic destinations are products of meetings and combinations of diverging modern and late-modern ideas for economic development, though ventures occasionally have a pre-modern character. What this book has shown is that destinations have to reorient their activities as tourism patterns change, but also that places reorient as they develop into tourism destinations.

Several chapters in the book examine the transformations of places into tourism destinations. Førde (Chapter 9) argues that this is an awakening process in Andøy. Situated north of the Lofoten islands, Andøy has for twenty years provided whale safari activities, and the community now seems to look at tourism as a way of living. Both here and in Målselv, tourism is not an alternative to the traditional industries, but a supplement that indeed provides these places with more diversified industrial platforms. This seems to be a common thread running through the book; tourism is something that adds to the livelihoods of the communities involved, for example the witchcraft waving in Strandir, Iceland, or sled dog tourism in Finnmark, Norway. In several of the presented cases, people seem to appreciate the attention foreigners pay to their place or culture. As Abram claims in Chapter 4, this is also one of the elements socializing them to the tourism system. However, there are also cases in this book which demonstrate scepticism towards touristification. In Andøy, Førde (Chapter 9) has observed that many think their region should avoid becoming as touristified as their neighbour region the Lofoten Islands; in Målselv, many were against profiling their community as Snowman land; and in Strandir, Iceland, people want their community to stay small and want to preserve as it has been. Thus, tourism is often perceived as an external force that tends to challenge established structures, places, industries, identities and conceptions. Some of the cases in the book refer to situations in which tourism has a quite different role, as in Sariseklä and Kilpisjärvi in Finland (Chapter 11), where it is dominant industry with a well-established position and where it involves quite different challenges. Here tourism is an internal force that is taken for granted, including the pattern in which it is performed. Kajan and Saarinen see this as a challenge, as there is limited capacity and orientation towards innovation or path creation.

Most of the cases presented in this book are about small scale tourism. This is the dominant form in rural parts of the Arctic, with some exceptions. One exception is cruise tourism. In Chapter 10, Ola Sletvold discusses these operations as a Fordist production. This industry, as it is today, collaborates with the tourism industry in their ports of call, where tourists are offered escorted excursions. In the Lofoten Islands, Fordist cruise production has been transmitted to the tourist companies serving tourists ashore, with a more standardized and volume based production, and more external steering and control. The local tourism providers are aware of this, but so far they see advantages. Their operations and incomes are more predictable and possible to plan. In the study of sled dog tourism, Jæger and Viken (Chapter 8) observed a similar effect of the cruise tourism in Alta: it is said to be the reason for the recent success of several local operators. As these authors maintain, this has also given these companies a more solid platform for serving other types of tourists.

Relations between cruise industry actors and local companies, as discussed in Chapter 10, are asymmetric: huge international and national companies dealing with small and local enterprises. There is also a national ground-handler company involved, which not only monopolizes contact between the two parties, but also buys local companies and merges them into a centrally controlled unit. This has happened to the guide services in the Lofoten Islands and in other places. However, concerning the land excursions, the model basically remains that local firms receive cruise tourists and entertain them but, as Sletvold shows, with a flavour of standardization. But the experience production is often based on personal or local knowledge and resources, and is not suited for a remotely controlled management. As Sletvold shows in Chapter 10, several of the small operators in the Lofoten Islands are aware of the risks and still claim their autonomy. Often these firms are founded by people making a living out of a preferred lifestyle. There also seems to be a development pattern, where smaller companies emerge and grow in the wake of the bigger ones (Carter 1991). This is also the case with whale safaris in Andenes and the pattern observed for the accommodation sector in the North Cape area (Viken and Aarsæther 2013).

The discussions of adaptation to tourism, and of industrialization and standardization processes, entangle the discussion of the postmodern consumption and position of an emerging culture economy. Several of the changes observed in this book can be seen as adaptation and contributions to such trends. The best examples are of course the witchcraft museum in Strandir, in Iceland, the Norwegian *Blånisseland* (Blue Elves Land) in Målselv, and the whale safari in Andenes, which also includes a land-based attraction. Sled dog tourism in Finnmark and snowmobile tourism in Kilpisjärvi can also be seen as materialization of the same trends (Ritzer 1999). The Blue Elves Land can also be seen as a sign of

'Disneyfication' (Ritzer and Liska 1998; Ritzer 1999). But many of the forms of tourism exemplified in the book are also local leisure activities, part of an 'aestheticized' everyday life (Featherstone 1992), and often enabled by tourism projects and ventures.

Destination Development as Politics

With reference to Foucault's articulation of power, we started out by quoting Cheong and Miller who state that 'power is everywhere in tourism' (2000, 372). This has partly been reflected in chapters using stakeholder and network approaches to the analysis of destination development. The stakeholder perspective is explicitly applied to tourism development in Strandir (Chapter 6), Nunavut and Svalbard (Chapter 14), Målselv (Chapter 9), and to cruise tourism in the Lofoten Islands (Chapter 7 and 10). The omnipresence of politics and power is obvious; there are relations and negotiations between stakeholders in most cases, and the authorities are often important parties.

Politics is also about steering, governing, establishing regimes, and management. In the book there are examples showing quite diverse ways in which this is done. In fact, tourism in most places is just one of many business sectors and a matter of general regulations. However, as shown in many cases, the authorities can have an active role, but in different ways. In Strandir, Iceland, the authorities established a public museum that has been a destination energizer. In Målselv, the municipal authority took on the role of developer, partner, and a sort of advocate for the company in charge of tourism development. In Muonio, Finland, the authorities have made plans and directed the framework for tourism development, whereas in Svalbard the authorities have the role as environment protector. Thus, four different forms of public intervention are observed: the authorities act as *doers* (or actors) – undertaking the factual development; act as *facilitators* – making way for a tourism project; perform as *steerers* – presenting frameworks for the development of tourism; and as *controllers*, watching the steps of the tourism industry. On the other side, the book also has examples of politics as non-intervention. In northern Sweden, the authorities have not done much to make tourism a more viable industry, and in Nunavut the authorities seems to have had a relaxed or laissez-faire attitude towards tourism. There is also a Norwegian case, Andøy, where public involvement has been legendary. In general, it seems like the authorities are most active when there is a particular need for new activities or employment. Therefore, tourism politics is not for the good times, but more a matter of public crisis management, as Førde sees it (Chapter 9). This is not necessarily the best premise for success.

There are three chapters in this book dealing with the political aspects of tourism destination development. One is Chapter 13, where Dieter Müller and Patrick Brouuder analyse tourism development in Jokmokk, Norrbotn, in northern Sweden, in relation to authority involvement. This is in fact almost insignificant, and the development observed has taken place on an industry and market basis. Thus, when a DMO (Destination Management Organisation) structure is established, it is an example of reactive policy – in fact, a parallel of what can be observed in Nunavut. Yet, Müller and Brouder show that there is a positive trend concerning tourism in the region, which is also appreciated by the authorities and which verbally supports ongoing development. In Chapter 14, Arvid Viken, Margaret Johnston, Torill Nyseth and Jackie Dawson compare tourism development in Nunavut, Canada and Svalbard, Norway. The focus is on responsible governance, as accountability and as responsiveness; the first, the old bureaucratic model, the other, modern network and adaptive governance. These examples show that, if accountability is chosen as a strategy, a cry for responsiveness might occur, as in Svalbard, but also, if responsiveness is the applied model, there is a risk of a passive public policy, as in Nunavut. A responsive governance philosophy is not particularly efficient if the parties involved are unable to make compromises. Therefore, it is argued, models of governance should be combined, giving more room to accountability and hierarchical steering in situations where there exist solutions that are obviously and objectively better than others, and relying on responsiveness when societal robustness is most important.

From Knowledge to Reflexivity

To generate knowledge in a field is a way of increasing awareness of its inherent social processes and structures. This is done in this book, through a series of case studies. From these we have learned about the ways in which tourism destinations have or have not come about, in Strandir, Iceland, in Målselv, Norway, and Nunavut, Canada; about the obstacles and challenges for those developing tourism, as in Munio, Kilpisjärvi and Sariselkä, Finland, and in Jokmokk, Sweden; and about encounters and relations between a manifold of actors involved in tourism, as in Lofoten, Andøy, Målselv and Svalbard, Norway, and in Nunavut, Canada. However, just as important as such knowledge is the growing insight concerning knowledge – its appearance, positions and production – in tourism destination development.

At the start of this epilogue, this question was addressed as an aim for the book: how does a contribution like this provide a reflexive turn within tourism discourses? As mentioned Kögler's (1997) exploration of the roots of self-consciousness can be used to enlighten some reflexivity aspects. According to him, there are three levels of reflexivity. The first he calls instrumental reflexivity, within which values and goals are taken for granted and the reflexivity is related to ways of realization. This involves being aware of the best ways to develop a tourism destination or resort. The three empirical parts of the book is basically providing a basis for such reflexivity. However, as shown in several chapters in this book, for instance in Målselv, Norway, well established knowledge has not been taken into use, and in the Finnish winter resorts Kilpisjärvi and Sarisälke, path dependency is observed as another type of deficiency concerning destination development. Thus, to put it simply, the knowledge base or instrumental reflexivity is not always as good as it probably should be.

The process of thinking about values and wider goals concerning one's practice, Kögler calls practical reflexivity. There are several cases in the book where such perspectives are present and even creating antagonism. For instance, concerning tourism development in East Svalbard, stakeholders disagree on how the area should be managed. All agree that the area should be preserved, but not on whether it can at the same time be presented for the public. A similar problem is observed in Munio, north Finland, where a plan to develop a hotel-based resort has been stopped due to environmental issues. Furthermore, in several of the chapters, the question is posed as to which values are supported or suppressed when growth is appraised. In fact, the book presents a series of examples of antagonism derived from the prevailing growth paradigm, explained in Chapter 2.

Thirdly, according to Kögler, a critical reflexivity exists, where questions are posed about the more profound values we serve and the goals on which we act. What kind of social or economic system do we support in dealing with tourism? '[S]ocial agents disclose certain aspects for social life according to socially inculcated interpretive schemas' (Kögler 1997, 225) and implicitly conceal others. This is to pose critical questions regarding our society as a whole; for instance, the capitalist economy, western hegemony, and globalization, which are addressed in the first section of the book. These issue are most concretely illustrated in the case study of East Svalbard. Is this an area where human beings need to perform? Is it not possible to reserve such places for polar bears, walruses and other wildlife? This way of looking at reflexivity also represents a way of looking at theoretical approaches: to what degree do researchers take their own approach as granted. Should, for instance, modern network based and responsive governance be a legitimate demand in areas without local inhabitants, as in East Svalbard? In order to develop tourism in a sustainable, responsible and resilient way, reflexivity is required. Subsequently, such big and ethical important questions should also be posed.

Conclusion: From Hegemonic Discourse to Situated Destination Knowledge?

Part of the reason for choosing a case study approach is a view of this as an appropriate way for understanding destination development as sociocultural and political processes. Methodological stances that consider other ways of unfolding such a field scarcely exist (Flyvbjerg 2006). However, this approach is also based on a view that knowledge is situational, circumstantial or contextual, related to the research situation and the role and position of the researchers. As mentioned above, to situate knowledge can be a way of being reflexive (Alvesson et al. 2008). This means that the studies represent knowledge about something going on here and now, seen through the eyes of a particular researcher, entering into a particular discourse and part of a certain research community. Situating knowledge is to recognize 'science as culture and the idea of science as a social construction' (Engelstad and Gerrard 2005, 3), and is 'concerned with conversations between different communities ... that ... are not only possible but also productive, stimulating, inspiring, and that ... will lead to better knowledge production'. (ibid., 5). Treating texts as situated knowledge, is placed in context by Merrifield (1995, 51): 'A situated understanding ... provides a position from which to organize, conceptualize, and judge the world. Yet it is always partial, never finished nor whole ... There are always different and contrasting ways of knowing the world, equally partial and equally contestable'. 'Total' theories do not exist, and a study always represents a part adding to the whole. Thus, part of the platform for this book is that 'true', 'correct' or 'objective' theories of tourism destination do not exist, but that there is a variety of ways in which this term and phenomenon should be understood. Together, this provides a rich platform both for analysing and developing tourism destinations.

The situatedness of the knowledge produced in this book is designed to contrast the hegemony that exists within this discursive field, dominated by the growth paradigm outlined in chapter two, and discussed in several others. This book shows how small scale tourism development based on lifestyle ventures such as those in north Norway and in Iceland; and how practical and experience based knowledge, more than academic knowledge, has paved the way for a ski resort in northern Norway. In several of these cases, the importance of authority involvement has been demonstrated. As partners and facilitators, the municipalities have enabled tourism to become part of a multiple industrial landscape in Andøy and in Målselv in Norway, and in Strandir in Iceland. Further, it is shown that mainstream market-oriented destination management, in northern Finland for example, is challenged by changing contexts such as those related to climate change. It is shown that the modern governance philosophy, based on networks and responsiveness, is challenged both by old management systems, as in Svalbard, and by a field that has not matured for self-governance in Nunavut. All this refers to developments where general ideas of growth, market analyses, strategic considerations, and business entrepreneurs have been of less importance. However, the book also contains chapters showing how well known business mantra as standardization (in Lofoten) and traditional market strategies (in northern Finland) are developing. Such operations are also part of the tourism destination landscape, situated in a complex and hybrid terrain, where tourism destinations emerge and crumble, flourish and fail.

References

Alvesson, M., Hardy, C. and Harley, B. (2008), 'Reflecting on Reflexivity: Reflexive Textual Practices in Organisation and Management Theory', *Journal of Management Studies* 45:3, 480–501.

Alvesson, M. and Sköldberg, K. (2000), *Reflexive Sociology. New Vistas for Qualitative Research* (London: Sage Publications).

Bourdieu, P. (1986), 'The Forms of Capital', in Richardson (ed).

Bourdieu, P. and Wacquant, L.J.D. (1992), *An Invitation to Reflexive Sociology* (Cambridge: Polity Press).

Carter, E. (1991), *Sustainable Tourism in the Third World: Problems and Prospects.* Discussion Paper No. 3 (London: University of Reading).

Chang, T.C. (2000), 'Theming Cities, Taming Places: Insights from Singapore', *Geografiska Annaler Series B, Human Geography* 82:1, 35–54.

Cheong, S. and Miller, M. (2000), 'Power and Tourism: A Foucauldian Observation', *Annals of Tourism Research* 27:2, 371–390.

Engelstad, E. and Gerard, S. (2005), 'Challenging Situatedness', in Engelstad and Gerard (eds).

Engelstad, E. and Gerard, S. (eds) (2005), *Challenging Situatedness. Gender, Culture and Production of Knowledge* (Delft: Eburon Delft).

Featherstone, M. (1992), *Consumer Culture and Postmodernism* (London: Sage).

Flyvbjerg, B. (2006), 'Five Misunderstandings about Case-Study Research', *Qualitative Inquiry* 12:2, 219–245.

Fries, C.J. (2009), 'Bourdieu's Reflexive Sociology as a Theoretical Basis for Mixed Methods Research', Journal of Mixed Methods Research 3, 326–328.

Kögler, H.H. (1997), 'Reconceptualizing Reflexive Sociology: A Reply'. *Social Epistemology* 11:2, 225–250.

Lynch, M. (2000), 'Against Reflexivity as an Academic Virtue and Source of Privileged Knowledge', *Theory, Culture and Society* 17:3, 26–54.

Merrifield, A. (1995), 'Situated Knowledge through Exploration: Reflections on Bunge's "Geographical Expeditions"', *Antipode* 27:1, 49–70.

Richardson, J. (ed) (1986), *Handbook of Theory and Research for the Sociology of Education* (Westport, CT: Greenwood Press).

Ritzer, G. (1999), *Enchanting a Disenchanted World. Revolutionizing the Means of Consumption* (Thousand Oaks, CA: Pine Forge Press).

Ritzer, G. and Liska, A. (1998), '"McDisneyization' and 'Post-Tourism': Complementary Perspectives on Contemporary Tourism', in Rojek and Urry (eds).

Rojek, C. and Urry, J. (eds) (1998), *Touring Cultures: Transformations in Travel and Theory* (London: Routledge).

Saarinen, J. (2004), 'Destinations in Change. The Transformation Process of Tourist Destinations', *Tourist Studies* 4:2, 161–179.

Tribe, J. (2005), 'New Tourism Research', *Tourism Recreation Research* 30:2, 5–8.

Viken, A. and Aarsæther, N. (2013), 'Transforming an Iconic Attraction into a Diversified Destination: The Case of North Cape Tourism', *Scandinavian Journal of Hospitality and Tourism* 13:1, 38–54.

Index

www.ingramcontent.com/pod-product-compliance
Ingram Content Group UK Ltd.
Pitfield, Milton Keynes, MK11 3LW, UK
UKHW021619240425
457818UK00018B/647